electrical and
electronics drawing

THIRD EDITION

electrical and electronics drawing

CHARLES J. BAER
Professor of Engineering
The University of Kansas

McGRAW-HILL BOOK COMPANY

New York	Kuala Lumpur	Panama
St. Louis	London	Rio de Janeiro
San Francisco	Mexico	Singapore
Düsseldorf	Montreal	Sydney
Johannesburg	New Delhi	Toronto

Library of Congress Cataloging in Publication Data
Baer, Charles J.
Electrical and electronics drawing.

   Bibliography: p.
   1. Electric drafting.   2. Electronic drawing.
I. Title.
TK431.B3 1972   *604'.26'213*   *78-38811*
ISBN 0-07-003008-1

Electrical & Electronics Drawing

   6 7 8 9   KPKP   7 9 8 7 6

The editors for this book were Cary Baker and Cynthia
Newby, the designer was Marsha Cohen, and its
production was supervised by James Lee. It was set
in News Gothic by York Graphic Services, Inc.
It was printed and bound by the Kingsport Press.

Contents

Preface

This book has been written both as a textbook and as a reference book. It is a textbook because it is organized in blocks of subject matter in logical sequence and has questions and problems at the end of each chapter. By answering the questions and working the problems in each chapter, the future technician or engineer can increase his proficiency in making electrical drawings. He may also expand his knowledge of electronic functions and systems somewhat, through exposure to the material presented herein.

This book is a reference book because it is quite comprehensive. It treats the main areas of the electrical industry, such as electronics, automation, microelectronics, electric power, and architectural wiring. It also includes large portions of many important standards (Appendixes B and C) as well as a glossary of terms (Appendix A) and Bibliography. Most of the standards which appear wholly or in part, or are similar to those which are printed here, are:

ANSI Y32.2   Graphical Symbols for Electrical and Electronic Diagrams

Mil Std 15-1   Graphical Symbols for Electrical and Electronic Diagrams (now obsolete)

ANSI Y32.9   Graphical Electrical Wiring Symbols for Architectural and Electrical Layout Drawings

Mil Std 15-3   Electrical Wiring Symbols for Architectural and Electrical Layout Drawings

ANSI Y32.14   Graphic Symbols for Logic Diagrams

ANSI Y14.15   Electrical and Electronic Diagrams

ANSI Y15.1   Illustrations for Publication and Projection

ANSI Z10.5   Letter Symbols for Electrical Quantities
Mil Std 12B   Abbreviations for Use on Drawings
Mil Std 16B   Electrical and Electronic Reference Designations
ANSI C1   National Electrical Code

In addition to the above-named standards, the professional draftsman may have to concern himself with some others. United States and Military Standards are available that pertain to such things as color coding, markings of assemblies, enclosure requirements, printed circuits, and testing. These are not included in this text.

As the reader will quickly see, the book is based largely on standards which promote standard usage in electrical drawings. The two main sources of standards are the American National Standards Institute and the United States Government,[1] but there are other organizations. The reader may obtain lists of the available standards by writing the organizations listed below:

ASME   American Society of Mechanical Engineers, 345 East 47th Street, New York, N.Y. 10017

American National Standards Institute, 1430 Broadway, New York, N.Y. 10018

EIA   Electronic Industries Association, 11 West 42nd Street, New York, N.Y. 10036

IEEE   Institute of Electrical and Electronics Engineers, 345 East 47th Street, New York, N.Y. 10017

JIC   Joint Industry Conference, Rockham Building, Detroit, Mich.

NMTBA   National Machine Tool Builders' Association, 2139 Wisconsin Avenue, Washington, D.C.

Superintendent of Documents, U.S. Government Printing Office, Washington, D.C.

The third edition of this book contains much of the material which was in the second edition. However, there have been

---

[1]Some of the American National and Military Standards are identical; for example, ANSI Y32.2 and Mil Std 15-1 were very nearly identical. But there are some areas, such as testing and marking, where only one standard (usually the Military) is available. Occasionally there is an area, such as abbreviations and identification for motor controllers, that is covered by a standard such as the Joint Industry Conference would print, which is not treated in much detail in any American National or Military Standard.

enough changes in electronics and related fields to make the previous edition obsolete. The same outline and chapters have been maintained in the belief that the majority of students have been well served by this format. Some students who use this text have had some electronics instruction but no drawing experience. Chapters 1 and 2 are for them. Other students have had formal drawing instruction but no electronics study. They will not need to read the first two chapters but should expend more time on control and semiconductor functions which are treated in later chapters.

The author wishes to make known his gratitude to the officials and engineers of the many companies and firms who were interested and thoughtful enough to assemble typical drawings for use in the book. He has endeavored to print these drawings as nearly as possible as they were originally drawn, although, in many cases, he was able to show only parts of the original drawings because of lack of space.

Charles J. Baer

*electrical and
electronics drawing*

Chapter 1
Techniques and lettering

Because drawing for the electrical and electronics industry includes so many types of drawings, it utilizes all the various graphical techniques in one place or another, at one time or another. In the beginning stages of some projects, rough freehand sketches may be the most appropriate way to show ideas graphically. On the other hand, the final stages of some projects may require precision drafting that necessitates the use of sophisticated and expensive equipment.

This book will not attempt to instruct the reader in the uses of these devices and methods. The bibliography lists some excellent textbooks on engineering graphics (engineering drawing). If the reader is not reasonably proficient in the use of conventional drawing instruments (triangles, T square, irregular curve, compass, and lettering guide), he may wish to consult one of these texts. Such books are often found in school libraries and sometimes in city libraries. These and other books, at both the high school and college levels, are probably adequately written to enable a person to teach himself how to use the more commonly known instruments well enough to work 95 percent of the problems in this book. The chances are, however, that he will not be able to do as fine work or as rapid work as students who have had one or more courses in mechanical drawing or engineering drawing.

As for the use of some of the more sophisticated drawing equipment, even the texts mentioned above do not attempt to do much. Very little, if any, mention is made of such de-

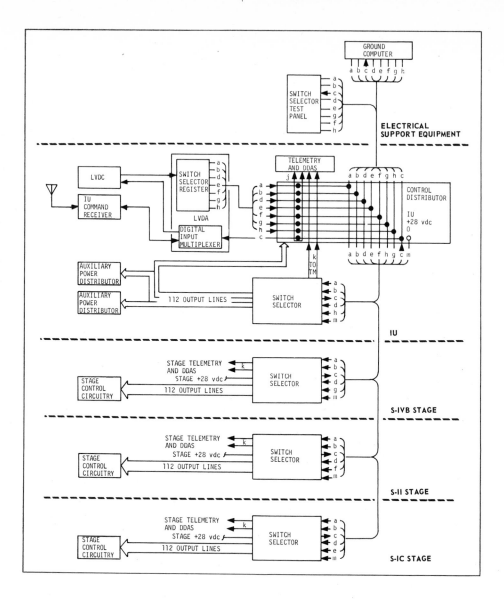

Fig. 1·1 *Saturn–Apollo switch functional diagram.*

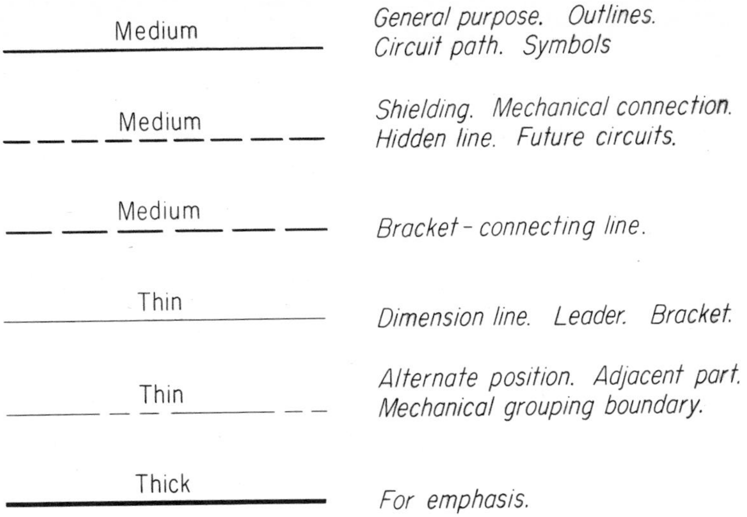

————— Medium —————	*General purpose. Outlines.* *Circuit path. Symbols*
— — — — Medium — — —	*Shielding. Mechanical connection.* *Hidden line. Future circuits.*
— — Medium — — —	*Bracket-connecting line.*
——— Thin ———	*Dimension line. Leader. Bracket.*
——— Thin ——— — —	*Alternate position. Adjacent part.* *Mechanical grouping boundary.*
▬▬▬ Thick ▬▬▬	*For emphasis.*

Fig. 1·2 Different lines and their uses in mechanical and electrical drawings.

vices or methods as photodrawing, coordinatographs, automatic drafting machines, and negative scribing, for example. The only ways in which the reader can become familiar with such devices are (1) to work with the equipment or watch experienced persons using the equipment, and (2) to read about them in technical journals such as *Graphic Science,* the *Journal of Engineering Design Graphics,* or *Electronics.*

1·1 Line work

Most electrical drawings require the drawing of many parallel lines, often horizontal and vertical. In certain cases these lines are drawn in black ink. In many cases, including work preliminary to ink drawing, the lines are drawn in pencil, with one weight of line predominating. Figure 1·2 shows the different weights used in making different types of lines. Sometimes it is difficult to get much difference in weights of lines made with a pencil, but this figure should give the reader an idea of what we strive for in the matter of lines.

There are quite a number of ways in which one can draw parallel lines, depending upon the accuracy required and the equipment available. Such methods are:

Fig. 1·3   Two styles of drafting machines. (Keuffel & Esser Co.)

1   With a plotting device such as the coordinatograph
2   With a drafting machine
3   With a parallel straightedge
4   With special section-lining devices
5   With a T square (and triangles)
6   With two triangles
7   With tape appliqués
8   Freehand

In general, the cost of equipment and the accuracy obtained decrease as we go from 1 to 8 in the above list. Figures 1·3 to 1·5 show some of the methods listed. For small drawings, a second triangle can be substituted for the T square, although this is not as convenient as using the latter.

This text will not discuss the making of inked drawings, a topic well treated by most of the standard drawing texts.

1·2 Circles and other shapes

Circles are generally made with the compass or with a template. Because circles in most electrical drawings are smaller than an inch, templates are used more frequently than the compass for circle construction. Several models of templates

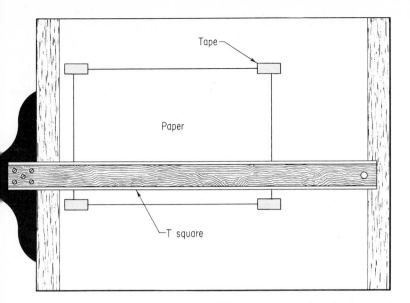

Fig. 1·4   Positioning of a small-size sheet of drawing paper on a drawing board.

are designed for the purpose of making circles only; others are for making circles, hexagons, squares, and triangles; and still others are for making electrical symbols.

Several concepts should be followed in using a template:

1   In order to draw a circle at the precise location desired, construction center lines (at least as long as the diameter

Fig. 1·5   Using a T square and triangle to draw vertical lines. (From Frank Zozzora, "Engineering Drawing," 2d ed., McGraw-Hill Book Company, New York, 1958. Used by permission.)

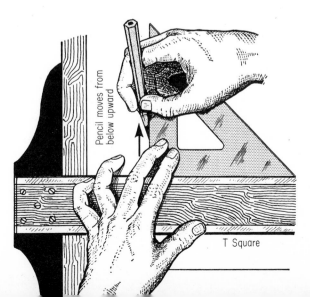

of the circle) must be put on the drawing before the template itself is positioned for the drawing of the circle.

2 For proper alignment, the template must be held firmly against a T square, drafting-machine arm, or some other suitable (horizontal) straight edge.

3 The pencil must be held *vertically* as the figure is being drawn. This is the only way to be certain of correct alignment and shape of figure. It may also be necessary to have a slightly duller point than usual on the pencil in order to achieve the desired line thickness of the figure.

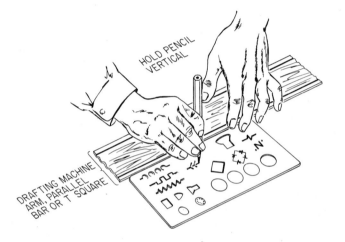

Fig. 1·6 *Using a template.*

Templates are very convenient for making electrical symbols for drawings. Care must be used in selecting from the many models that are on the market because the templates are fairly expensive and some may not have the appropriate symbols or the correct sizes of symbols for certain types of electrical drawings.

1·3 The importance of lettering

Electrical drawings utilize all the conventions and shortcuts of engineering drawings in order to present a graphical concept of devices or systems. However, the line work, symbols, and other patterns of a drawing are not sufficient to give the complete picture. Considerable lettering is required on most electrical and electronics drawings. Figure 1·1 is a typical drawing for showing the amount and kind of lettering required on an electrical drawing. Some drawings require even more in proportion to the amount of line work than this drawing. In many cases, as much time is used in doing the lettering on the drawing as is used in doing the line work. In drawings which are to be used for the manufacture, construction, or installation of equipment, lettering is usually done freehand. In drawings that are to be used over and over again or in those which for certain reasons are to be printed in books, manuals, or journals, lettering is usually drawn with mechanical lettering devices.

In every case, lettering must be good. Poor lettering not only ruins the appearance of a drawing that is otherwise good but improves the chances of costly mistakes being made in the reading of the drawing. It so happens that the lettering in Fig. 1·1 was done mechanically, mainly because it was a part of a printed maintenance manual. But before this final drawing was made, another one using freehand lettering had been drawn.

1·4 Types of engineering lettering

Practically all the alphabets used in technical drawings are single-stroke Commercial Gothic. Four types of these single-stroke letters in use today are:

1 Vertical uppercase (capitals)
2 Inclined uppercase
3 Inclined lowercase ("small" letters)
4 Vertical lowercase

The alphabets above are arranged in what the author believes is the order in which each type is used in United States industry, with type 1 being used the most, type 2 the next, etc. One who wishes to become expert in the making of electrical or electronics drawings should be proficient in doing freehand lettering of types 1, 2, and 3.

The person who has had no lettering instruction and who wishes to become proficient at freehand lettering should practice one type of alphabet at a time. A logical sequence of alphabets would be (1) vertical uppercase, (2) inclined uppercase, and (3) inclined lowercase. With each alphabet the

TYPE 1

ABCDEFGHIJKLMNOPQRST
UVWXYZ &
1234567890 $\frac{1}{2}$ $\frac{3}{4}$ $\frac{5}{8}$

TYPE 4

ABCDEFGHIJKLMNOPQRSTUVWXYZ &
1234567890 $\frac{1}{2}$ $\frac{3}{4}$ $\frac{5}{8}$ $\frac{7}{16}$
FOR BILLS OF MATERIAL, DIMENSIONS
& GENERAL NOTES

TYPE 6

abcdefghijklmnopqrstuvwxyz
Type 6 may be used in place of
Type 4, for Bills of Material and
Notes on Body of Drawing.

Fig. 1·7 *Typical American Standard alphabets.*

student should develop a set of numbers that appear to match or fit with the letters. As a rule, numbers are usually made narrower than the letters of the companion alphabet. Thus, the number zero should be narrower than the letter O of the same alphabet.

1·5 Standard alphabets

Lettering for drawings is standard throughout the United States. Three standardized alphabets are shown in Fig. 1·7. There may be slight differences between alphabets authorized by one agency and another, and also between alphabets shown in drawing textbooks, but these differences are minor. The shapes of the letters are, in general, the same.

Figure 1·8 shows vertical uppercase letters placed in blocks that are six small squares high. Borrowed from a fa-

Fig. 1·8 *Proportions and suggested strokings for letters. (From Thomas E. French and Charles J. Vierck, "Fundamentals of Engineering Drawing," McGraw-Hill Book Company, New York, 1960. Used by permission.)*

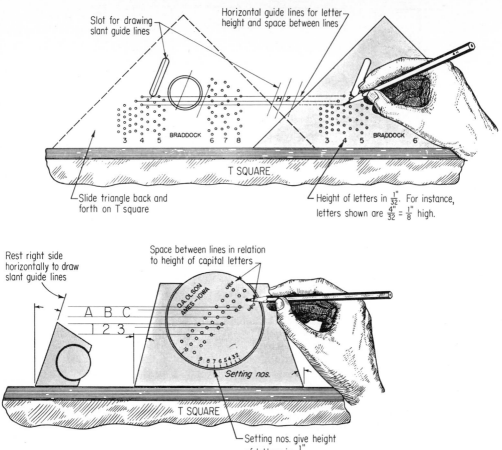

Fig. 1·9 Use of lettering guideline devices. (From Frank Zozzora, "Engineering Drawing," 2d ed., McGraw-Hill Book Company, New York, 1958. Used by permission.)

mous textbook, this illustration shows the comparative widths of all letters and the suggested order of strokes for right-handed persons. This order does not have to be followed, but it is used by many draftsmen. Left-handers will probably have to develop their own order of stroking, using whatever is most comfortable and effective.

1·6 Guidelines

Practically all freehand lettering must make use of guidelines. These are very light, thin lines that locate the tops and bottoms of capital letters, numbers, etc. They should be just dark enough to see, because erasing them is not practicable after letters are penciled in; therefore they should be as nearly invisible as possible.

There are a number of ways to draw guidelines. The most convenient way is to use a device especially made for that purpose. Two such devices are shown in Fig. 1·9. In this illustration, guidelines for capital letters have been constructed with both the Braddock and the Ames lettering devices. In the lower figure, lines have also been partially drawn for the bodies of lowercase letters but were stopped near the letter C and number 3. Guidelines may also be made with T squares and triangles in conjunction with scales.

Different combinations of letters and figures require different arrangements of guidelines. If only uppercase letters are to be used, just the top and bottom guidelines are necessary (see Fig. 1·10a). For lowercase letters, the intermediate, or "waist," line must be used with the other two lines, as shown in Fig. 1·10b. A "drop" line may be placed below the baseline to facilitate making the descenders. This is not done very often, probably because the normal spacing provided by lettering devices does not allow enough space for drop lines. Occasionally it is desirable to have a line exactly centered between the upper and lower guidelines. This would be helpful, for example, when many fractions are part of a note (see Fig. 1·10c). Because fraction heights are usually made twice those of whole numbers and capital letters, additional lines may be added above and below, as shown at the right end of

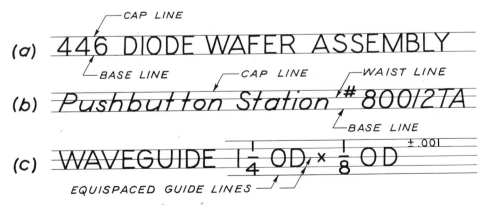

Fig. 1·10 *Use of guidelines for different situations.*

Fig. 1·10c. This would produce a series of lines all equally spaced. This type of spacing can be obtained with the Ames lettering instrument for nine different sizes of letters and with some models of the Braddock-Rowe lettering triangle for one size of letter.

Another way to align the tops and bottoms of letters is to use guidelines already drawn on a piece of paper. This paper is positioned under the sheet of tracing paper or transparent plastic material on which the drawing is to be made. The guidelines show through the drawing medium and are followed just as if they were drawn on the medium itself.

1·7 Parts lists and tables

Figure 1·11 shows a part of a typical parts list, or bill of material. In practice, the heavy horizontal dividing lines are usually spaced ¼, ⁵⁄₁₆, or ⅜ in. apart, and the letters are made

ITEM	REFERENCE	DESCRIPTION	MAT'L
1	310-19506	CHASSIS	
2	35A-19472	BRACKET, MOUNTING	
3	DD-6040A	CAPACITOR, 1 MF 25 WVDC	
4	50C-19503	CABLE ASSEMBLY	
5	30103744	LOCK SHAFT	CRS
6	60104321	SCREW #4-40 x $\frac{1}{4}$	STEEL

Fig. 1·11 Layout for a parts list with suggested vertical spacing.

not more than one-half as high as this vertical spacing. One convenient way to organize such a list is to use equally spaced guidelines, as shown in the lower part of Fig. 1·11. For instance, if the lines are spaced ¹⁄₁₆ in. apart (using a lettering guide set for No. 4 letters, as in Fig. 1·11), letters and numbers will be ⅛ in. high and centered between the horizontal dividing lines, which will be ¼ in. apart. Abbreviations are frequently used, examples being CRS for cold-rolled

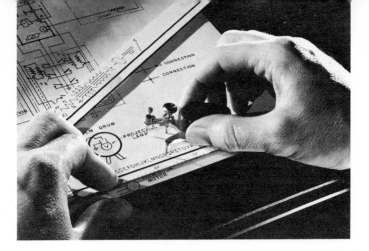

Fig. 1·12 A mechanical lettering device. Leroy lettering template and scriber. (Leroy is a registered trademark of Keuffel & Esser Co., by whose courtesy the above photograph is shown.)

steel and AL for aluminum. Capital letters are almost always used. Parts lists and other tabular arrangements often accompany, or are part of, electrical drawings. Their exact arrangements (formats) are not standardized.

1·8 Mechanical lettering devices

A number of devices for making engineering letters mechanically have been successfully tried and used in the drawing room. They fall into two classes: (1) the stencil, or incised-letter, type, with which the draftsman himself puts the letters on the drawing medium, and (2) the special typewriter, with which a typist puts the letters on the drawing after the draftsman or engineer has written or otherwise indicated what material he wants typed. Examples of the first type are shown in Figs. 1·12 and 1·13. These can be used for penciled or inked

Fig. 1·13 Another mechanical lettering device. (This Rapidesign lettering template is sold by Gramercy Guild Group, Inc., by whose courtesy the above photograph is shown.)

letters. The fountain-pen type of lettering pen (the Rapido-graph, for example) can be used with the stencil type such as that shown in Fig. 1·13.

Although mechanical lettering devices have been on the market for more than 30 years, they have not replaced free-hand lettering on many engineering or electrical drawings. For this reason, the drafting or engineering student cannot expect to get along in his profession without being able to do neat, proficient freehand lettering.

SUMMARY   The techniques of making electrical drawings are the same as those used in making other engineering drawings. With proper study, practice, and equipment, nearly any person can make most types of electrical drawings. Some types, however, require expensive precision equipment which is not described in most instructional books or courses.

A good knowledge of the shapes of letters belonging to several standard alphabets is required of the person who is engaged in drawing electrical or electronics drawings. One can learn to letter well by practicing fundamental strokes and imitating good samples of lettering. In the foreseeable future there is little likelihood that mechanical lettering or computers will replace freehand lettering in most electrical drawings.

QUESTIONS   1·1   Name five different ways of drawing parallel horizontal lines.

1·2   Why is it necessary to have different line widths and different types of lines in electronics drawings?

1·3   For what purposes are templates used in the drawing room?

1·4   In addition to achieving uniform height of capital letters, in what other respects do we strive to achieve uniformity?

1·5   What are some advantages of uppercase lettering? Of lowercase?

1·6   Are the letters of an alphabet generally narrower or

wider than the numbers which accompany the same alphabet?

1·7 Why can poor lettering on a drawing be costly?

1·8 Why do some agencies and publishers require mechanical lettering?

1·9 Show, by sketching, an effective way of making guidelines for a parts list.

1·10 How does the shape of a curve differ between inclined uppercase and vertical uppercase letters? (Sketch or describe.)

1·11 What is a drop line?

1·12 Name or list several ways to draw circles.

1·13 In general, how many types of lettering are used in United States industry?

PROBLEMS LETTERING EXERCISES

1·1 In ⅛-in. uppercase letters, letter the following terms or sentences:
a. Focus coil, red
b. Horizontal drive
c. P-353446 detector
d. Waveguide assembly
e. Yaw guidance cptr.
f. 110-Hz detector
g. Center-accessory compartment

1·2 With ⅛-in. uppercase letters, letter the following:
a. Left-aft cabin light
b. μA 741
c. P-11C425 waveguide screw
d. KS-19087 L-13 connector detail
e. All resistors are ¼ watt $\pm$ 5%

1·3 With No. 3 or 4 inclined lowercase letters, letter the following notes:
a. 12.47-KV switchgear location
b. Wheel-pulse input pickup: master warning panel

c. $C1$, $C2$, $C3$ are for connections of power-factor-correcting capacitors

d. Tinned leads .040 ±.003 dia.

1·4 In a space about 2 × 4 in., make a set of general notes as follows: Resistance value in ohms. Capacitance values less than one in MF, one and above in PF, unless otherwise noted. Direction of arrows at controls indicates clockwise rotation. Voltages should hold within ±20% with 117 V a-c supply.

1·5 With uppercase letters ³⁄₃₂ or ⅛-in. high, letter the following specific notes:

a. Knockout for 1½-in. conduit may enter top or bottom of cabinet.

b. ⁹⁄₁₆ DIA (4 holes) for wall mtg.

c. 5½ × 6 cutouts for sec. cable conn.

d. Tie-bolt holes ½ max.

e. All pins 0.093 ±0.003 DIA

f. .187 ±.003 DIA, 4 pins

1·6 Make a parts list with the following headings: Item No., Part No., Description, No. Required, Remarks. Place the following items in the appropriate boxes, 1 through 6.

1	21—4	100-ohm resistor	2	
2	31—14	.001 UF capacitor	4	Disc
3	40—66	Antenna coil	1	
4	431—10	Terminal strip	4	3-lug
5	511—4	Transistor	1	2N155
6	19—27	Control with switch	1	1 Megohm

1·7 Letter a lighting-fixture schedule using the following headings: Mark, Manufacturer, Cat. No., Mounting, Watts, Lamps, Finish. Place the following items in the appropriate boxes, A through E.

A Skylite, 50, Recessed, 300W, Silver Bowl, Std.

B Lightcraft, 60R, Ceiling, 150W, Par 40, Satin Alum.

C Prescolite, 1313–6630, Recessed, 300W, 1F, Std.

D Prescolite, 1015–6615, Recessed, 150W, 1F, Std.

E Prescolite, 90–L, Recessed, 150W, 1F, Std.

1·8 Make a schedule, using No. 4 or No. 5 letters, for a

printed wiring board. Use the block format which is used with parts lists.

hole no.	X-coord.	Y-coord.	position tolerance	diameter
11	.486	2.120	.028	.304/.314
12	1.988	2.100	.028	.304/.314
13	1.520	2.120	.028	.304/.314
14	.838	1.712	.028	.063/.068
15	1.738	1.590	.014	.090/.097

.063/.068 hole to be plated through

1·9 Using inclined lowercase letters, letter the following equipment descriptions, one below the other, as they would appear beside a vertical one-line diagram:

a. 1/0 ACSR

b. 3-30KV Line-type Lightning Arresters

c. 23KV 200 AMP S&C Hook-operated Disconnect

d. 3000KVA SC-3917KVA FAC West. Transformer 22000△ − 4160Y / 2400 volts

e. 3-CT's 1000/5

f. Watthour Meter W/Demand

g. 5KV 600A Air Circuit Breaker

h. 400 AMP 5KV Enclosed Disconnect @ 3–4/0 cu.

1·10 Using No. 3 or 4 vertical uppercase letters, letter the following statements:

a. Unless otherwise specified, resistance values are in ohms and capacitance values are in picofarads.

b. DC coil resistance values under one ohm are not shown on the schematic diagram.

c. Arrows on controls indicate clockwise rotation.

d. Control viewed from shaft end

e. Saturable-coil line reactors

1·11 Using No. 3, or 0.10-in. high, vertical uppercase letters, letter the following graph terms:

a. Discriminator output, volts DC

b. Base current, milliamperes

c. DC collector voltage, volts

d. Frequency response curve of amplifier

e. Duration of fades in seconds

f. E_k in electrons per ion

g. No. of television sets produced, millions

1·12 Using a mechanical lettering device or a lettering template, letter the terms shown in Prob. 1·9 or 1·10 in ink.

1·13 Using $\frac{5}{32}$ or $\frac{6}{32}$-in. uppercase letters, letter the following titles, centering them from left to right on the sheet:

a. Fidelity curve of W4C amplifier

b. Schematic diagram of Model 130 television receiver

c. One-line diagram of Littlefield substation

1·14 Letter the titles shown in Prob. 1·13, using a mechanical lettering device or template.

1·15 Make a title plate for the drawings you have made in this course. Details concerning the size of the plate and the exact information to be lettered on the plate will be supplied by the instructor.

1·16 Lay out and letter a block title for one or more drawings of this course. The information which should be included in the title block will be supplied by the instructor.

MECHANICAL DRAWING EXERCISES

1·17 Make a drawing of the laser diagram shown in Fig. 1·14. Drawn to the outline dimensions shown, it will fit on a sheet of 8½ × 11 paper.

1·18 Make a drawing of the photodiode shown in Fig. 1·15. Use 8½ × 11 paper. Include lettering. (May be combined with Prob. 1·25 or 1·26.)

1·19 Make a drawing of the third-rail detail shown in Fig. 1·16. Some of the dimensions are supplied. Make your drawing to similar proportions so that it will fit nicely on 8½ × 11 paper.

1·20 Make a drawing of the end view of a satellite oscillator assembly to the dimensions shown in Fig. 1·17.

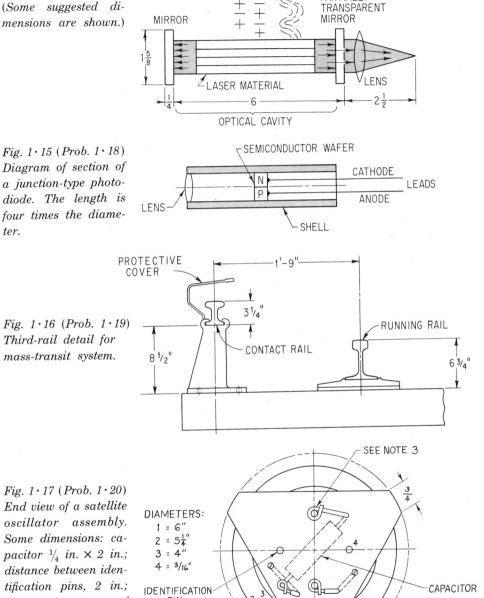

Fig. 1·14 (Prob. 1·17) Schematic diagram of laser block diagram. (Some suggested dimensions are shown.)

PUMP

MIRROR

PARTIALLY
TRANSPARENT
MIRROR

$1\frac{5}{8}$

LASER MATERIAL

LENS

$\frac{1}{4}$

6

$2\frac{1}{2}$

OPTICAL CAVITY

Fig. 1·15 (Prob. 1·18) Diagram of section of a junction-type photodiode. The length is four times the diameter.

SEMICONDUCTOR WAFER

CATHODE

LEADS

ANODE

N
P

LENS

SHELL

Fig. 1·16 (Prob. 1·19) Third-rail detail for mass-transit system.

PROTECTIVE
COVER

1'-9"

$3\frac{1}{4}$"

RUNNING RAIL

$8\frac{1}{2}$"

CONTACT RAIL

$6\frac{3}{4}$"

Fig. 1·17 (Prob. 1·20) End view of a satellite oscillator assembly. Some dimensions: capacitor $\frac{1}{4}$ in. × 2 in.; distance between identification pins, 2 in.; between centers of large loops, 2 in.

SEE NOTE 3

$\frac{3}{4}$

DIAMETERS:
1 = 6"
2 = $5\frac{1}{4}$"
3 = 4"
4 = $\frac{3}{16}$"

IDENTIFICATION
PIN

4

CAPACITOR

2 3
1

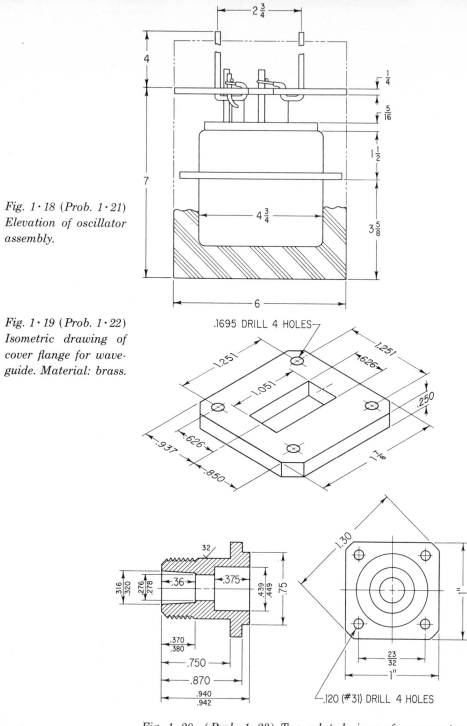

Fig. 1·18 (Prob. 1·21) Elevation of oscillator assembly.

Fig. 1·19 (Prob. 1·22) Isometric drawing of cover flange for waveguide. Material: brass.

Fig. 1·20 (Prob. 1·23) Two related views of a connector. The threads are $^5/_8$–24.

Include lettering. Use 8½ × 11 paper. (May be combined with the next problem on a larger sheet.)

1·21 Make a drawing of the front view of the oscillator assembly shown in Fig. 1·18. Use 8½ × 11 paper. Suggested scale: full size.

1·22 Make a two-view (top and front or top and side) drawing of the waveguide cover flange shown in Fig. 1·19. Dimensions may be converted to common fractions (see conversion table in Appendix B) or to two-place decimals. Suggested scale: twice actual size. (This will fit on 8½ × 11 paper.) This may be dimensioned, if your instructor so indicates.

1·23 Draw two views of the connector shown in Fig. 1·20. Draw to twice the actual size (2:1). Show the dimensions that are given in the illustration. For drawing purposes, convert dimensions to common fractions (see Appendix B) or two-place decimals. Use 8½ × 11 paper.

1·24 Make outline drawings of the semiconductors shown
1·25 in Figs. 1·21, 1·22, and 1·23, respectively. Use an
1·26 enlarged scale so that one or two of these objects (two views of each, as shown) will fit on an 8½ × 11 drawing sheet. Many of the dimensions are supplied

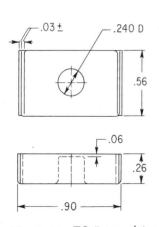

Fig. 1·21   TO-5 transistor.
(See pictorial view, Fig. 2·18.)

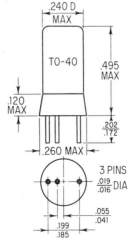

Fig. 1·22   Transistor.

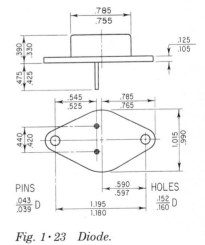

Fig. 1·23   Diode.

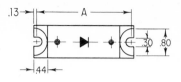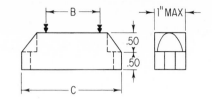

Fig. 1·24   Diode.

as *limits;* that is, the upper and lower limits of each dimension are shown. Select the average of the two limits for your drawing dimension, and round it off to the nearest hundredth (or sixteenth) of an inch. Optional additions to drawing the outline are:

a. Showing the dimensions on the drawing

b. Making a specification drawing which includes:

   (1) The outlines

   (2) Manufacturer's name and model number

   (3) A list of specifications such as contact resistance, dielectric strength, ampere rating, life expectancy, and other information

   (4) The dimensions

   (5) Component symbol

If it is desired to make a specification drawing, the above-listed information will have to be obtained from the instructor or from a manufacturer's transistor manual.

1·27   Make an outline drawing (two or three views) of the diode shown in Fig. 1 · 24. Use an enlarged scale and show the dimensions. Then underneath, or to the right side of your drawing, letter the following table or schedule. Use 8½ × 11 paper.

DIMENSIONS IN INCHES			
type	*A*	*B*	*C*
CR101	2⅛	1⅛	2⅜
CR102	2⅛	1⅛	2⅜
CR103	2⅛	1⅛	2½
CR105	3¼	1¾	3½
CR110	5½	4	5¾

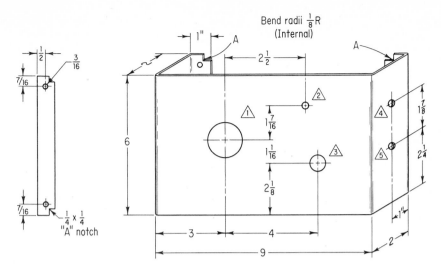

Fig. 1·25 (Prob. 1·28) Pictorial view of subchassis.

1·28   Make two or three views (front, top, side) of the sub-
        chassis shown in Fig. 1·25. May be shown as a
        development. Show the dimensions. Holes are to be
        drilled. Their diameters are: No. 1, — 1½; No. 2, — ¼;
        No. 3, — ⅝; No. 4, — ¼; and No. 5, — ¼. Material:
        SAE 1020 steel. Add notes: Break all sharp edges;
        remove all burrs; degrease per 51606; bright dip per
        51606; dry per 51606.

1·29   Make a drawing of the fuel cell shown in Fig. 1·26.

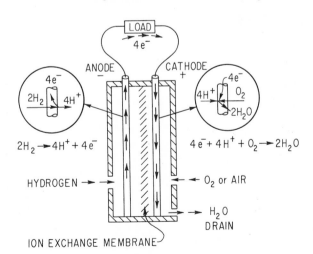

Fig. 1·26 Schematic of Apollo fuel cell.

Chapter 2
Pictorial drawing

As the electronics and electrical fields continue their phenomenal expansion, more and more devices that require pictorial representation, either in place of or as adjuncts to standard orthographic projection, are encountered. Also, electrical drawings are being read by more and more technical personnel who have not had training in the reading and making of engineering drawings. These are two factors which have been responsible for the increase in the amount of pictorial drawing being done in the electrical industry.

Good examples of pictorial drawing used in the electrical field are those used by our armed forces maintenance personnel, drawings used by assembly-line workers for assembling electronic equipment, and those used by companies who manufacture do-it-yourself kits. Figure 2·1 is a typical example of the latter.

The types of pictorial drawings most often found in the electrical field are:

1 Isometric drawing
2 Oblique drawing
3 Dimetric drawing
4 Perspective drawing

Brief treatment of the first three categories of pictorial drawing will follow. References will be given for the more complicated subject of perspective drawings.

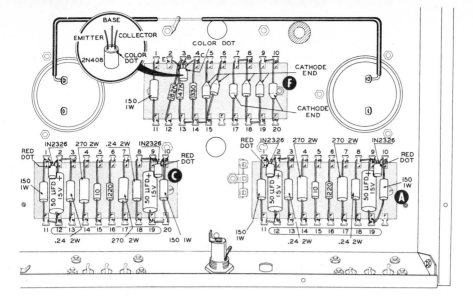

Fig. 2·1   *Pictorial drawing used for the assembly of electronics equipment. (Heath Company.)*

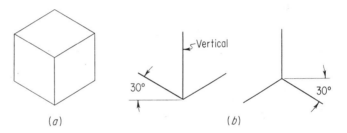

(a)   (b)

Fig. 2·2   *An isometric drawing of a cube, and the isometric axes. (a) Cube. (b) Two arrangements of isometric axes.*

2·1 Isometric drawing[1]

To draw an object at such an angle or position that all edges will be foreshortened equally is conveniently possible. For example, all lines of the cube shown in Fig. 2·2 have been drawn to the same length; that is, they have been foreshortened equally. This drawing, called an isometric drawing or isometric projection, was obtained by using three axes, as shown in Fig. 2·2b. Rectangular-shaped objects lend themselves readily to this type of drawing because all lines, or edges, are drawn along or parallel to the three isometric axes, and can be scaled directly.

[1]Liberal translation of the word *isometric* from the Greek gives "iso" meaning "the same" and "metric" meaning "measurement."

A simple and direct way to make an isometric drawing of an object composed of rectangular-shaped surfaces is to draw an outline of a cube and then to measure off distances from edges of this cube to the corners and edges of the object. Such step-by-step procedure is shown in Fig. 2·3b, the completed drawing appearing in Fig. 2·3c.

Isometric drawings may be made with instruments (the 30–60 triangle is a natural choice), or they may be drawn freehand. If an instrument drawing is to be made, any suitable scale may be used, and only the one scale selected will be used in making measurements along all three axes. Isometric sketches are sometimes shaded to give additional clarity. Hidden lines are not usually shown. Dimensions may be added if necessary.

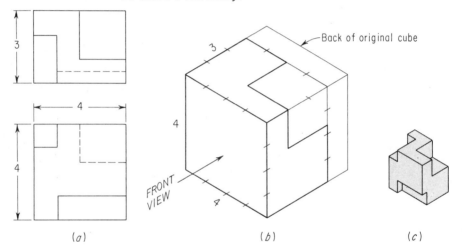

(a)   (b)   (c)

Fig. 2·3 Preparation of an isometric drawing begun as a 4-in. cube. (a) Front and top views (given) of the object. (b) Starting to draw the notch in the rear, upper right part. (c) Completed isometric drawing.

2·2 Examples of isometric drawing

Figure 2·4 shows isometric drawings of a flatpack integrated-circuit package and a full-wave bridge. For these particular objects, one pictorial drawing may be more descriptive and informative than two or three orthographic (principal) views. An isometric drawing of a console for use in the automatic assembly of circuit boards appears in Fig. 2·5.

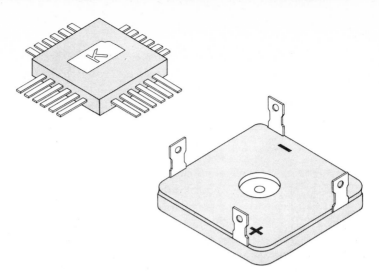

Fig. 2·4   Isometric drawings of a flatpack IC and a full-wave bridge.

2·3 Oblique drawing

Another type of pictorial representation which is easily made is oblique drawing, or oblique projection. In this type of drawing, one face of the object is drawn in its true shape and the other visible faces are shown by parallel lines, or projectors, drawn at the same angle (usually 30° or 45°) with the hori-

Fig. 2·5   Isometric drawing of a console for an automatic circuit-board assembler.

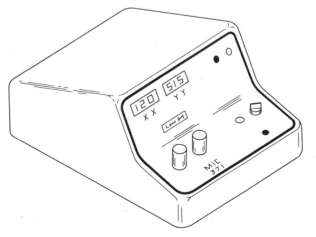

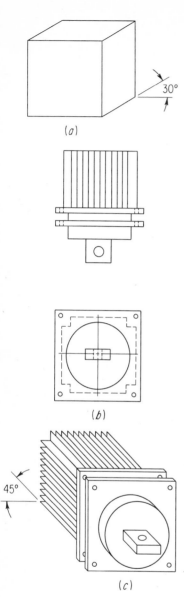

(a)

(b)

(c)

Fig. 2·6 Oblique drawings of a cube and silicon rectifier. (a) Cube. (b) Front and top views. (c) Oblique drawing of the rectifier.

zontal. Figure 2·6 shows two objects drawn in this manner.

The projectors of the cube (Fig. 2·6a) are drawn to the right at a 30° angle and are shortened to preserve the appearance of the cube. Such foreshortening of lines along the oblique axis is called *cabinet drawing,* and more often than not the projectors are drawn to one-half the size of the lines they represent. The rectifier (Fig. 2·6b) is suitable for oblique drawing for two reasons: (1) The face drawn perpendicular to the line of sight (in the plane of the paper) contains circular parts which can be drawn with a compass or circle template, and (2) the other axis is comparatively short, thus making it unnecessary to shorten the projectors. The projectors of the rectifier were drawn 45° to the left and are not foreshortened. If the rectifier were drawn in the isometric manner, they would have been drawn at a 30° angle and all circles and arcs would have to be drawn as ellipses or parts of ellipses. Like isometric drawings, oblique drawings may be made freehand or with instruments.

2·4 Examples of oblique drawings

Figure 2·7 shows an oblique drawing of a transistor with radiator. Figure 2·8 is an oblique drawing of the same console which appeared in Fig. 2·5.

2·5 Dimetric projection

Another form of pictorial representation, similar to isometric drawing, is dimetric drawing, or dimetric projection. In this type of drawing, the reference cube is viewed from such an angle that two edges are equally foreshortened. The third axis (the

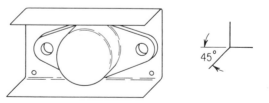

Fig. 2·7 Oblique drawing of a transistor with a heat-radiation attachment.

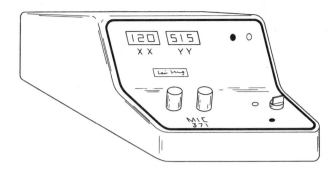

Fig. 2·8 Oblique drawing of a console. See Fig. 2·5 for comparison purposes.

lines marked 3 in Fig. 2·9a) is shortened a different amount, however, and therefore cannot be measured with the same scale that is used for the other two axes. Examples of dimetric drawing are shown in Fig. 2·9. Dimetric drawing is not as popular as isometric and oblique drawing for two reasons: (1) Difficulty is encountered in drawing circular parts as represented dimetrically, and (2) it is necessary to use two different scales when laying out a dimetric drawing. However, a dimetric drawing will produce a less distorted, more pleasing effect than will either of the other two types of drawing.

2·6 Perspective drawing

The three methods of pictorial representation described thus far in this chapter produce only approximate representations of objects as they appear to the eye. Each type produces varying degrees of distortion of any device or system so drawn. However, because of the ease and quickness of their execu-

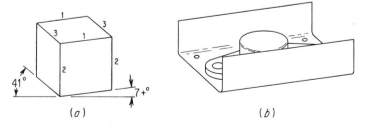

Fig. 2·9 Dimetric drawing. (a) Cube. (b) TO-66 transistor can with radiator.

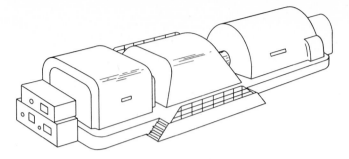

Fig. 2·10   Perspective drawing of a steam turbine generator.

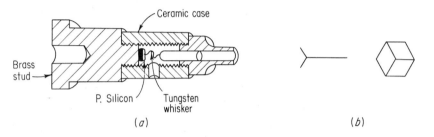

(a)   *(b)*

Fig. 2·11   (a) An isometric section of a diode. (b) Showing how the isometric axes have been oriented.

tion, they are the types of pictorial drawing most often used in the electrical industry.

On certain occasions, the exact pictorial representation of an object as it actually appears to the eye may be necessary or desirable. To do this requires that the principles of perspective drawing be observed. An example of such a drawing is shown in Fig. 2·10.

Perspective drawing is a time-consuming process which requires more explanation than can conveniently be given in this book. Most texts on engineering drawing (see Bibliography) cover this subject.

2·7 Pictorial sections

Occasionally, the interior detail of a device can be appropriately shown by means of a pictorial section. All types of pictorial views—isometric, oblique, and perspective—can be used to show unusual interior details with full sections, half sections, or broken-out sections.

Figure 2·11 is a section view showing the construction of a typical silicon point-contact diode. It is a full isometric section, in which the main axis is horizontally oriented. This is in contrast to the other isometric drawings in this chapter in which this axis has been drawn in a vertical position.

The same figure also illustrates the fact that standard section lining (crosshatching) symbols are not always adequate for electrical and electronics drawings. There is no standard section symbol for silicon. The silicon slice in this case was not section lined. Each material has been labeled—a necessary, or at least highly desirable, feature when the American Standard section symbols are not followed.

2·8 Hidden lines and center lines

Hidden lines and center lines are not usually drawn in pictorial drawings. However, hidden lines may be shown if special hidden details are desired to be shown, and center lines may be required if an object is to be dimensioned.

2·9 References

More detailed explanation of the construction required for oblique, isometric, and dimetric views may be found in the reference books listed in the bibliography at the end of the book.

SUMMARY The use of pictorial drawings has increased greatly in the last decade. This is true for all technical areas and in particular for the electronics field. Most of the well-known types of pictorial drawing have been used. Probably the two types that are the easiest to make are isometric and oblique drawings. The choice of which to use could well depend upon the shape of the object to be represented pictorially. Isometric, dimetric, and oblique drawings do not quite show the true shape of any object so drawn, thus some distortion results. If this distortion is objectionable, then one must use perspective drawing—which is considerably more difficult and therefore more expensive. Pictorial sections are often desirable. Hidden lines, section lines, and center lines as a rule are not shown. However, they may be placed on a pictorial drawing if the occasion so demands.

2·1 Why are pictorial drawings becoming increasingly more numerous in the electrical and electronics fields?

2·2 How many triangles would be required in making a small isometric drawing?

2·3 How many triangles would be required in making a small oblique drawing?

2·4 Is a cabinet drawing a refinement of an isometric drawing or of an oblique drawing?

2·5 Name a situation where there would be an advantage in making an oblique drawing instead of an isometric drawing.

2·6 Name a situation where making an isometric drawing would have an advantage over making an oblique drawing.

2·7 What are the angles most often used in drawing "projectors" in oblique drawings? Are these angles measured from the horizontal or from the vertical?

2·8 Name two requirements that make a dimetric drawing more difficult to execute than an oblique drawing.

2·9 What disadvantage does a perspective drawing have when compared to an isometric drawing? To an oblique drawing?

2·10 Some pictorial drawings contain all three of the following: center lines, dimension lines, and hidden lines. Is this statement correct?

PROBLEMS In many of these problems, one or more types of pictorial drawing can be used to depict the object or system satisfactorily. In only a few cases, therefore, has a specific type of pictorial view been required. Many of the objects can be drawn pictorially as freehand sketches or as mechanical drawings. Your instructor may, in addition, require you to put dimensions on the pictorial drawing.

2·1 Make a pictorial drawing of the bracket shown in Fig. 2·12. Diameters of the holes are 0.15 and 0.30 in. Position the holes as closely as you can to the way they appear in Fig. 2·12. Use 8½ × 11 paper.

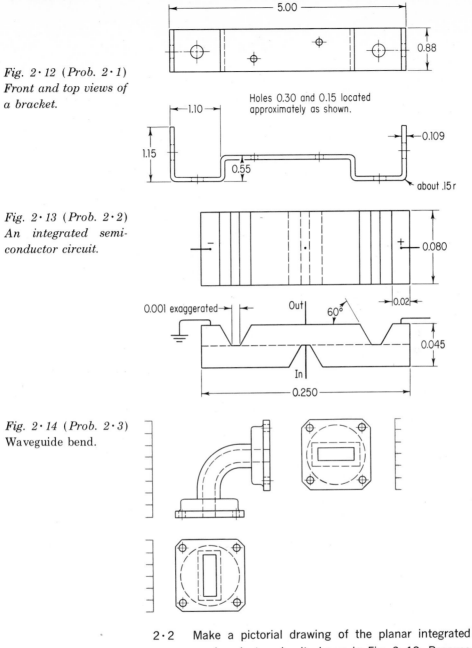

Fig. 2·12 (Prob. 2·1) Front and top views of a bracket.

5.00

0.88

Holes 0.30 and 0.15 located approximately as shown.

1.10

1.15

0.55

0.109

about .15 r

Fig. 2·13 (Prob. 2·2) An integrated semiconductor circuit.

−

+

0.080

0.001 exaggerated

Out

60°

0.02

0.045

In

0.250

Fig. 2·14 (Prob. 2·3) Waveguide bend.

2·2   Make a pictorial drawing of the planar integrated semiconductor circuit shown in Fig. 2·13. Because the object is so very small, it will have to be drawn many times (perhaps 20 or 30) its actual size. Use 8½ × 11 paper.

2·3   Make a pictorial drawing of the microcircuit waveguide accessory shown in Fig. 2·14. Use the rough scales along the left edge for correct proportions. Use 8½ × 11 paper.

2·4 Make a pictorial drawing of the printed-circuit connector shown in Fig. 2·15. Use the graduations along the bottom and at the left to achieve the correct proportions. Use 8½ × 11 paper.

2·5 Draw a pictorial section view of the phototransistor shown in Fig. 2·16. Estimate those sizes or distances which are not dimensioned in the book. Use 8½ × 11 paper.

2·6 Draw a pictorial view of one or both objects shown in Fig. 2·17. These will have to be drawn many times their actual size. Use 8½ × 11 paper.

2·7 Draw a pictorial view of the transistor shown in Fig. 2·18. This will have to be enlarged if a good accurate drawing is to be made. Make an isometric or oblique pictorial, or both, as assigned by your instructor. Use 8½ × 11 paper.

2·8 Make a pictorial drawing of the silicon diode rectifier of Fig. 2·19. Use an enlarged scale. Use 8½ × 11 paper.

2·9 Make a pictorial drawing of the transistor shown in Fig. 2·20. It should be enlarged for drawing purposes. Use 8½ × 11 paper.

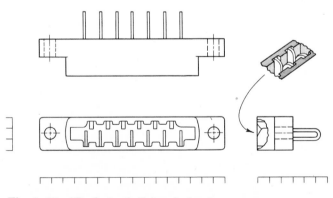

Fig. 2·15 (Prob. 2·4) Printed-circuit connector.

Fig. 2·16 (Prob. 2·5) Phototransistor details.

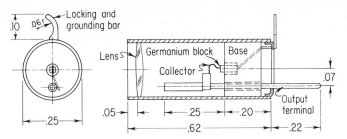

Fig. 2·17 (Prob. 2·6) Plastic TO-3 (left) and Triac (right).

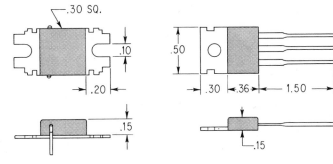

Fig. 2·18 (Prob. 2·7) TO-5 transistor with radiator.

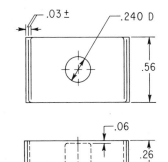
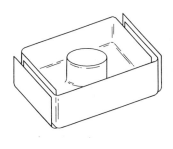

Fig. 2·19 (Prob. 2·8) Silicon rectifier.

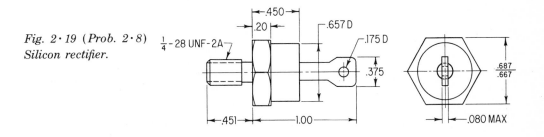

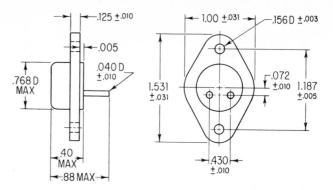

Fig. 2·20   (Prob. 2·9) Power transistor.

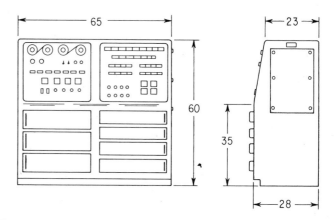

Fig. 2·21 (Prob. 2·10)   Front and side views of Saturn IC checkout console.

2·10   Make a pictorial view of the console shown in Fig. 2·21. This may be an isometric or oblique drawing. Use 8½ × 11 paper.

2·11   Make a complete pictorial diagrammatic drawing of the industrial television arrangement shown in Fig. 2·22. Label all items. Use 11 × 17 or 12 × 18 paper.

2·12   Construct a complete pictorial drawing of the automatic paging system shown in Fig. 2·23. The front panel of each of the six blocks is shown below. The rest of each unit can be shown as the plant PA system, upper left, appears. If oblique projection is

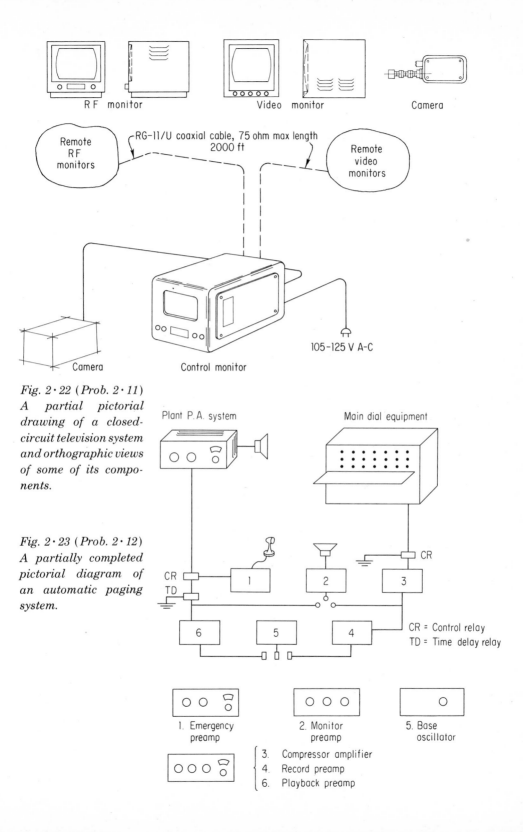

RF monitor Video monitor Camera

RG–11/U coaxial cable, 75 ohm max length 2000 ft

Remote RF monitors

Remote video monitors

105–125 V A-C

Camera Control monitor

Fig. 2·22 (Prob. 2·11) A partial pictorial drawing of a closed-circuit television system and orthographic views of some of its components.

Fig. 2·23 (Prob. 2·12) A partially completed pictorial diagram of an automatic paging system.

Plant P.A. system

Main dial equipment

CR

CR
TD

1 2 3

6 5 4

CR = Control relay
TD = Time delay relay

1. Emergency preamp

2. Monitor preamp

5. Base oscillator

3. Compressor amplifier
4. Record preamp
6. Playback preamp

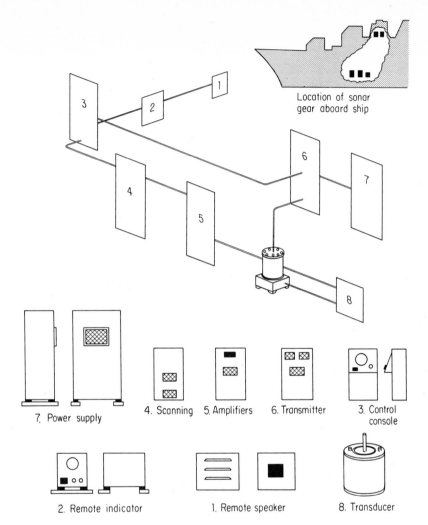

Location of sonar
gear aboard ship

7. Power supply   4. Scanning   5. Amplifiers   6. Transmitter   3. Control
console

2. Remote indicator   1. Remote speaker   8. Transducer

Fig. 2·24   (Prob. 2·13) Sonar system pictorial problem.

used, the emergency preamp (block 1) will appear almost the same as the plant PA, upper left. Label each unit and the relays. Use 11 × 17 or 12 × 18 paper.

2·13   Draw the sonar system in Fig. 2·24. Label each unit and show the schematic location as it is shown in the upper center. Label all parts. Use 11 × 17 or 12 × 18 paper.

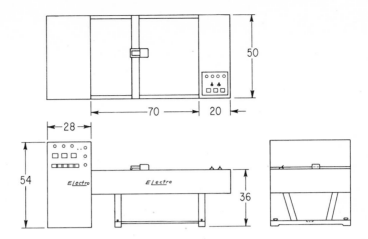

Fig. 2·25 (Prob. 2·14) *Outline drawing of an automatic drafting machine.*

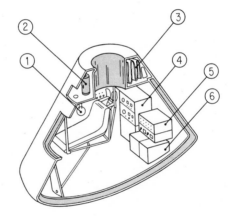

Fig. 2·26 (Prob. 2·15) *Pictorial section of Apollo Command module. (1) Oxygen surge tank. (2) Vacuum-cleaner storage. (3) Drogue parachutes. (4) Rate and attitude gyro. (5) Power servo assembly (above) and guidance computer (below). (6) CO_2 absorber cartridge. Ht. = 11.125 ft; dia. = 12.83 ft.*

2·14 Make a pictorial drawing of the automatic drafting machine shown in Fig. 2·25. Key dimensions have been shown. Use 11 × 17 or 12 × 18 paper.

2·15 Make a pictorial section of the Apollo command module shown in Fig. 2·26. Label all items.

Chapter 3
Device symbols

A very large portion of drawing in the electrical-electronics field is of a diagrammatic nature. This diagrammatic drawing makes great use of symbols. Originally, these symbols were drawn to look something like the parts they were to represent. However, the parts and the symbols have changed considerably through the years, until now there is not much physical resemblance between symbols and the respective parts. In some cases, though, a symbol may give a suggestion of the shape of the part it represents. Examples are the inductor, resistor, headset, antenna, and knife switch.

In order to facilitate making and reading electrical drawings, representatives of industry and government have come together to make up lists of standard symbols to be used in drawings. There are several sets; three important ones are the following:

ANSI Y32.2 Graphic Symbols for Electrical and Electronics Diagrams (also IEEE No. 15)

Mil Std 15–1 Graphical Symbols for Electrical and Electronic Diagrams (superceded by ANSI Y32.2)

IEC (International Electrotechnical Commission) Graphical Symbols

The ANSI and Military Standards have converged, and the IEC standard uses many of the symbols shown in the United States standards. In Appendix C are shown many of the symbols that appear in ANSI Y32.2. The reader should be able to identify readily and remember those pages where the symbols can be found. Some of the symbols most often used in electronics and electrical drawings will be shown and discussed in this chapter.

Symbols of common nonelectronic devices

3·1 The battery

One of the simplest symbols is that which represents a single-cell battery, shown in Fig. 3·1a. (The horizontal line represents the path of the signal, or current, and is not a part of the battery symbol itself.) The longer of the two lines always represents the positive terminal. The short line is about half the length of the long one. Multicell batteries can be shown with four or more lines (four is enough), as shown in Fig. 3·1b. Polarity symbols have been shown but theoretically are not necessary. However, these are sometimes used for emphasis or when it is believed the reader may be unsure. The third battery symbol shows two taps—one fixed and one (with arrow) adjustable.

3·2 The capacitor

Another often-used symbol that is easy to draw is the capacitor—sometimes called a condenser; its main function is to store electrical charge. The straight line can be made about the length of, or larger than, the long line of the battery symbol. The entire symbol is sometimes drawn twice the size of the single-cell battery symbol. (More about sizes will be found in Sec. 3·11.) The curved line represents the outside electrode in fixed-paper and ceramic-dielectric capacitors, the moving element in variable and adjustable types, and the low-potential element in feed-through capacitors. Showing the polarity sign

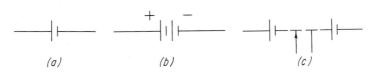

Fig. 3·1 Symbols for batteries. (a) Single cell. (b) Multicell with polarity marks added. (c) Multicell with two taps.

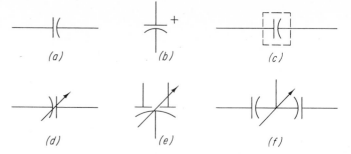

Fig. 3·2 Symbols for capacitors. (a) General. (b) Polarized. (c) Shielded. (d) Adjustable or variable. (e) Adjustable or variable differential. (f) Split-stator capacitor.

usually indicates an electrolytic capacitor. Good drawing practice should be observed in the drawing of the shielded symbol. Corners should be full and complete, and the dashed lines should not touch or intersect the signal path. In the case of Fig. 3·2e, the capacitance of one part increases as the capacitance of the other part decreases. But in the split-stator capacitor, the capacitances of both parts increase simultaneously.

Unfortunately, there are other symbols used for capacitors that do not appear in the American or Military Standards. The reader should be on the alert for these symbols. Because the international standard approves the symbol shown in Fig. 3·3a, we have shown the IEC standard for capacitors in Appendix C.

3·3 Chassis, ground, circuit return

It is usually necessary to connect parts of a circuit to a chassis, ground, frame, etc. If the conducting connection is to a chassis or frame that may have substantially higher potential than ground or the surrounding structure, the chassis symbol

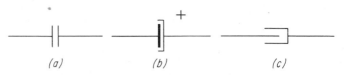

Fig. 3·3 Other symbols used for capacitors. (a) IEC standard. (b) Sometimes used for electrolytic capacitors. (c) Sometimes used in industrial electronics drawings.

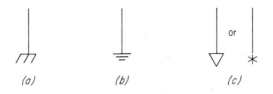
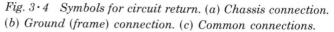

(a)   (b)   (c)

Fig. 3·4 Symbols for circuit return. (a) Chassis connection. (b) Ground (frame) connection. (c) Common connections.

(Fig. 3·4a) should be used. If the conducting connection is to earth, a body of water, or to a structure which serves the same function (such as a land, sea, or air frame), the symbol shown in Fig. 3·4b should be used. The common connection symbol should be used for common-return connections at the same potential level. This triangle symbol is used when a common conductor such as a ground bus or battery bus is used. Then, in the lower part of the diagram, a key is used in which the symbol is shown and the words GRD BUS or BAT BUS—24V or other appropriate terms are indicated. Sometimes a letter system is used if more than one type of common circuit return is shown on the same diagram. The triangle may be omitted and proper identification made where the asterisk appears on the right-hand symbol of Fig. 3·4c, although the triangle is more meaningful to the author than the end of a line without a symbol. (See Figs. 6·18 and 6·30.)

3·4 Connections and crossovers

Two systems are approved for showing connections (junctions) or crossovers. One is the dot system, as shown in Fig. 3·5a, b, and c. The other is the no-dot system, used in Fig. 3·5d to f. The no-dot system has some weaknesses, but

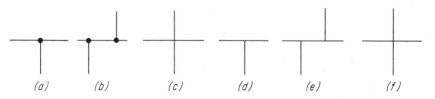

(a)   (b)   (c)   (d)   (e)   (f)

Fig. 3·5 Connections and crossovers. (a) and (b) Connections for dot system. (c) Crossover for dot system. (d) and (e) Connections when using no-dot system. (f) Crossover, no-dot system.

it is the preferred method, according to the most recent printing of ANSI Y14.15. But many prefer the dot system for clarity.

3·5 The inductor

The inductor, or induction coil, is used in a great many ways. In different situations it may be a transformer winding, a reactor, a radiofrequency coil, or a retardation coil. Although the American National Standard Institute approves both symbols shown in Fig. 3·6a, nearly all the examples in ANSI Y32.2

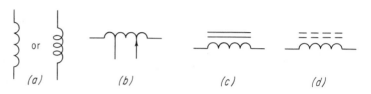

Fig. 3·6 Inductor symbols. (a) General symbols. (b) With fixed and variable taps. (c) With magnetic core. (d) With ceramic-type core.

and Y14.15 use only the symbol shown at the left. Many United States companies, however, still use the more complicated helical symbol. Because induction increases as the frequency increases, no cores (or air cores) are usually necessary at high frequencies. But at low frequencies, magnetic cores are often used to increase the induction. Ceramic cores are often composed of magnetic materials called *ferrites*.

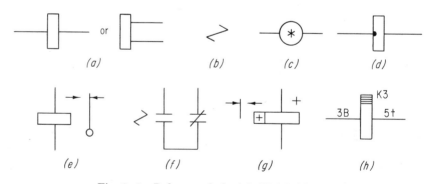

Fig. 3·7 Relay symbols. (a) (b) (c) (d) Approved relay coil symbols. (e) and (f) Relays with transfer contacts. (g) Polarized relay with transfer contact. (h) Slow-release relay with letter-number designations.

3·6 The relay coil

Three symbols are approved for the relay coil, also known as the solenoid. These are shown in Fig. 3·7a to c. The asterisk in Fig. 3·7c indicates that a letter or value, such as the relay number should be in the circle. The semicircular dot in Fig. 3·7d indicates the inner end of the winding. Figure 3·7e and f shows combinations of relay coils with switches; the whole combinations are called *contactors*. The switches shown here are often called *contacts* and are turned off or on by action of the relay coils. In Fig. 3·7f, the left-hand contact is normally open and the right-hand contact (with inclined line drawn at 60° to the horizontal) is normally closed. Action of the relay will close the left contact and open the right one. The term *slow-release* (Fig. 3·7h) is only relative. This relay closes one contact before it activates another, but the total action happens very quickly.

3·7 The resistor

Two approved symbols for the resistor are shown in Fig. 3·8a. The older, zigzag symbol is probably used the most. The rectangle symbol, originally used in the electrical-controls field, has been adopted by a number of companies in other areas. Made with a 60° angle between adjacent lines, the zigzag symbol needs only three points on each side, unless extra

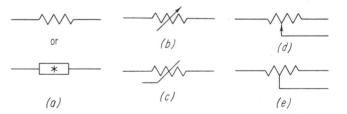

Fig. 3·8 Symbols for resistors. (a) General. (b) Variable or adjustable. (c) Non linear. (d) With adjustable contact. (e) Tapped.

taps or other special features require more. The asterisk within the rectangle means that identification should be placed within or near the rectangle. Typical values would be 210 (ohms), 20K (20 kilohms), or 1MEG (1,000,000 ohms).

Resistors may be fixed, variable (rheostat), or tapped—with either fixed or variable taps (see Fig. 3·8c and d). They are usually linear. If not, a special nonlinear symbol is available. Resistors are used for such purposes as dividing voltage, dropping voltage, developing heat, and minimizing current and voltage surges. The shaft of the arrow in Fig. 3·8b is drawn at about 45° in this and other symbols that require variability.

3·8 The switch

The purpose of the switch is to open or close circuits. The words *break* or *make* are often used instead of *open* and *close*. Mechanical-switch symbols are usually combinations of contact symbols, and they may be fixed, moving, sliding, nonlocking, etc. They are shown in the position in which no operating force is required. This is sometimes called the *normal*, or *initial*, position, in which the circuit is not energized. Figure 3·9f is not really a switch symbol; it simply defines the switching function. If, instead of a bar, an *X* were shown, we would have an open (make) contact. Many other devices—diodes, transistors, tubes, and cryotrons—also perform switching functions.

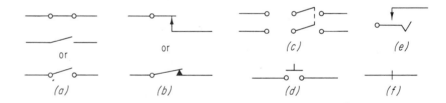

Fig. 3·9 Symbols for switches. (a) Single-throw, general: closed and open. (b) Nonlocking, momentary: opening (break). (c) Double-throw, two-pole. (d) Pushbutton: closing (make). (e) Locking, closing (make). (f) Switching-function symbol: closed contact (break).

Rotary-switch symbols are drawn looking at the knob or at the rear end and with the operational sequence in the clockwise direction. The projection on each segment of the "wafer" represents the moving contact. When several functions are performed, the tabular form of presenting information, as in Fig. 3·10, is preferred. Here, dashes link the

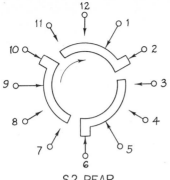

S2

POS.	FUNCTION	TERM.
1	OFF (SHOWN)	1-2, 5-6, 9-10
2	STAND BY	1-3, 5-7, 9-11
3	OPERATE	1-4, 5-8, 9-12

S2 REAR

Fig. 3·10   Position-function relationships for rotary switches. (*From American Standard Drafting Manual, ANSI Y14.15, "Electrical and Electronics Diagrams." Used by permission of the publisher, The American Society of Mechanical Engineers, 345 E. 47th St., New York, N. Y. 10017.*)

terminals that are connected. In position 2, for example, terminals 1 and 3 are connected (not terminal 2) and terminals 5 and 7 and 9 and 11 are connected.

3·9 The transformer

The symbols used for inductors are also utilized in the drawing of transformers. The American National Standard Y32.2 approves the use of either the helix symbol or the more rudimentary symbol. The international and obsolete military standards approve only the latter. We shall use the simpler symbol in most of our drawings. Transformers are made with air cores (usually found in high-frequency circuits) or with iron or laminated cores (found primarily in low-frequency

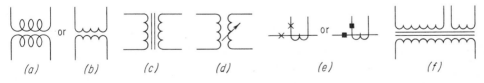

Fig. 3·11   Symbols for transformers. (*a*) and (*b*) General. (*c*) If it is desired to show a magnetic core. (*d*) With one winding having adjustable inductance. (*e*) Current transformer with polarity markings. (*f*) Power transformer.

a-c circuits). The standard, however, does not require that the two parallel lines be drawn for magnetic or metallic cores. The polarity symbols in Fig. 3·11e are placed so that the instantaneous direction of current into one polarity mark corresponds to current out of the other polarity mark.

The power transformer supplies power (usually in different amounts) to two or more circuits. The secondary windings are often drawn with different numbers of loops, or cusps, to suggest the relative voltages going to these circuits. Sometimes the secondary circuits have additional taps.

3·10 New device symbols

In an expanding area such as electronics, new devices for which no symbols exist will be forthcoming. Two methods of solving the problem of portraying these devices are in common use: (1) a new symbol for each device may be invented or designed, or (2) a rectangular block with appropriate identification may be inserted in the diagram at the place where the new device would be placed. Figure 3·12 shows a symbol which was designed by someone to show a cryogenic switch.

3·11 Size of symbols

Theoretically, size of a symbol is not important. But the relative sizes of symbols are an important matter. These relative sizes are shown in the standards, in Fig. 3·13, and in Appendix C of this book. Figure 3·13 suggests minimal and maximal sizes for several commonly used symbols. The minimum sizes are for small diagrams or for larger diagrams that are filled with many devices. The maximum sizes are for large drawings, say those on paper larger than 12 × 18 in. Other factors may dictate what size symbols should be drawn. If a student or draftsman is using a template to make symbols, he is pretty much limited to the sizes made by the template. Crowded conditions may require that he draw his symbols smaller than the suggested sizes. Lines are usually of medium weight—the same weight (width) used to draw the connecting paths and other lines in a drawing. However, symbols are sometimes made with heavy, thick lines for purposes of emphasis.

Fig. 3·12   A symbol invented to show a cryotron.

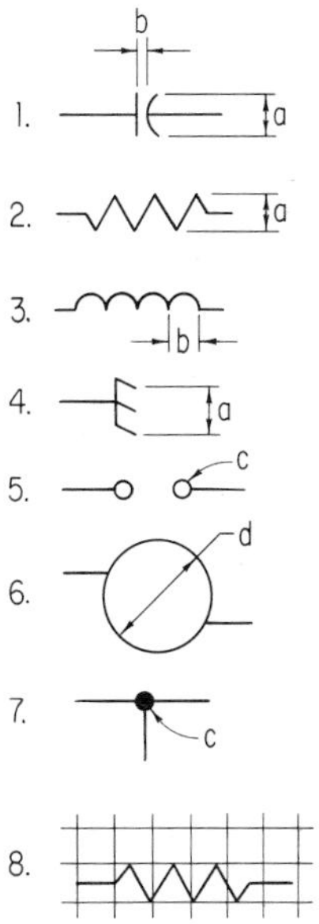

	Measurements in fractions of an inch						
	minimum				maximum		
	a	b	c	d	a	b	c	d
Capacitor	.25	.06			.40	.10		
Resistor	.15				.30			
Inductor		.15				.25		
Chassis	.25				.35			
Terminals			.06				.10	
Transistor envelope				.60				.80
Connection			.06				.12	

Fig. 3 · 13 Suggested sizes for electrical device symbols. (The lowest figure shows a resistor symbol that has been drawn on cross-ruled paper.)

3 · 12 Drafting aids

There are at least three ways in which symbols can be made more quickly than by drawing them with conventional drawing instruments. These methods use:

1 Templates
2 Preprinted symbol cutouts
3 Typesetting equipment with symbols

There are many designs of electrical templates on the market, and no one template will make all the symbols that are in use today. Great care should be used in the selection and purchase of such a device. A template should be used with a T square or drafting machine, as described in Chap. 1. Some templates are so thin that they tend to slip under the edge of the T square and hence are not entirely satisfactory. Some users prefer to raise the template off the paper. One method is to place the template over a triangle and use the

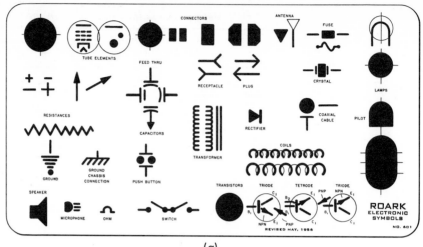

(a)

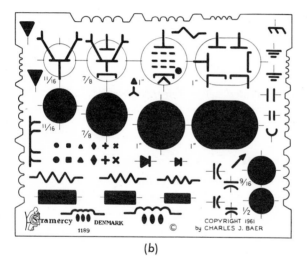

(b)

Fig. 3·14 Templates and applique's (pressure-sensitive adhesives). (a), (b) Templates (the black areas represent open spaces). (c) Appliques. (Tech Tac and Bishop Graphics, Inc.)

grooves that are over the open part of the triangle.

Preprinted symbols are manufactured as contact (pressure-sensitive) adhesives, or appliqués. These are cut out or lifted off a sheet of preprinted symbols and then positioned on the drawing in the correct place.

The typesetting machine is used for special types of electrical drawings. It has advantages where many drawings of great similarity are made within a company. Many engineers draw

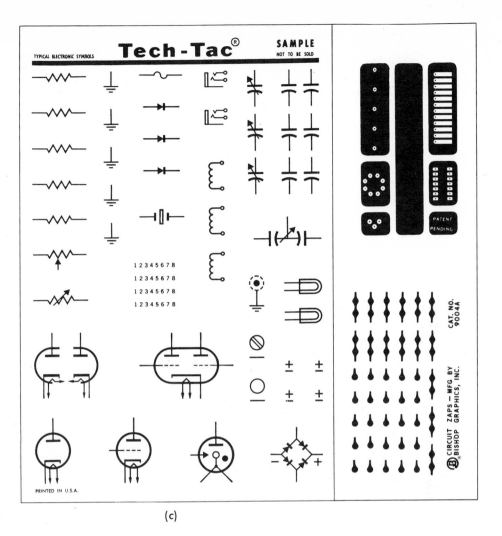

(c)

freehand diagrammatic sketches, and the final drawing is made on a typesetting machine or automatic drafting machine.

Symbols of electronic devices

3·13 The transistor

Called a *solid-state device*, the transistor is made of semiconducting material (germanium or silicon) which has characteristics that lie midway between those of good conductors (such as copper) and insulators (such as glass).

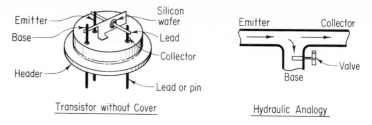

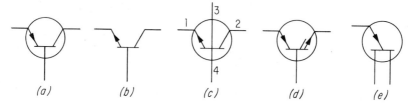

Transistor without Cover

Hydraulic Analogy

Fig. 3·15 *Assembly and action of triode transistor.*

(a) (b) (c) (d) (e)

Fig. 3·16 *Symbols for transistors. (a) PNP type. (b) NPN type. (c) Tetrode. (d) PNPN type. (e) Unijunction with P-type base.*

Many transistors have three leads, and thus are called *triodes*. The three outside connections are connected to the base, emitter, and collector. Figure 3·15 shows details of transistor assembly and a hydraulic analogy of the way a triode transistor performs. A change in the current flowing through the low-impedance base circuit causes a corresponding change in the current flowing to the collector circuit. The collector circuit, having high impedance, is the output circuit. Figure 3·16 shows symbols representing several types of transistors. Several things should be pointed out about transistor symbol construction:

1 The circle (envelope) does not have to be drawn if no confusion arises or if no leads are attached to the envelope.
2 Orientation, including a mirror-image presentation, does not change the symbol meaning.
3 Except for the unijunction transistor, the base symbol is drawn about one-third of the way "up" if the envelope circle is drawn.
4 Collector and emitter lines are drawn at about 60° to the base symbol, and the arrowheads do not touch the base line.

Although the envelope circle is not required, many engineers and draftsmen prefer, or recommend, that it be drawn. Figure 3·16*b* does not include the envelope. In Fig. 3·16*c*, the leads have been numbered, starting with the emitter, which is the standard order when numbering is desired. The student should become thoroughly familiar with the symbols and current-flow characteristics of PNP and NPN transistors. Figure 3·17 supplies this information. The circular symbol with ∼ inside indicates a signal source. We have only briefly touched upon the electron action of the transistor, a fascinating subject. There are several good books that go into this subject fully. Several are listed in the bibliography.

3·14 The diode

Another often-used semiconductor device is the diode. Figure 3·18 shows symbols for several different kinds or arrangements of diodes. The basic rectifier symbol is usually not encircled, but it—as well as the other symbols shown—may be enclosed in a circle if it is desired. This symbol points in

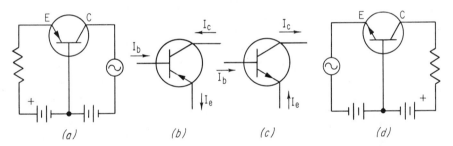

Fig. 3·17 Transistor biasing. (a) Biasing of PNP transistor in an amplifier circuit. (b) and (c) electron flow in PNP and NPN transistors. (d) Proper biasing of NPN transistor.

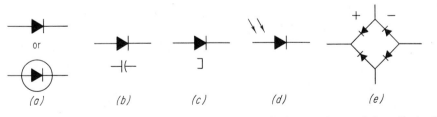

Fig. 3·18 Diode symbols. (a) General (rectifier). (b) Capacitance (varactor). (c) Tunnel diode. (d) Photodiode. (e) Bridge.

the direction of conventional current, sometimes called the "easy" direction. The bar part of the symbol corresponds to the cathode of an electron tube. Forward-biased diodes produce a "square" wave. Diodes are not always forward-biased, however. Diodes are used for different purposes such as rectification, detection, amplification, and switching. The tunnel diode, for example, has a negative-resistance characteristic that permits it to be used as an amplifier or as a switch. The bridge rectifier provides an easily smoothed d-c output with about twice as much average voltage as the full-wave bridge.

The location and installation of semiconductors requires the following procedures:

1 Don't exceed their maximum voltage, current, or power ratings.
2 Use adequate heat sinks for power devices.
3 Avoid prolonged exposure to heat during soldering.
4 Don't locate sensitive circuit devices adjacent to heat-producing power devices.
5 Don't solder semiconductors into an electrically live circuit.

3·15 The electron tube

A vacuum tube consists of an evacuated envelope, a cathode which supplies electrons, and an anode which collects the electrons; it may have one or more grids. (See Fig. 3·19.)

When the cathode is heated, it releases electrons, which flow to the anode (or the *plate*) when the latter has a positive potential with respect to the anode. A grid, which is a fine mesh or helix that offers little obstruction to the electron flow if it is at the same or slightly lower potential than the cathode, may be placed between cathode and anode. However, if this grid is attached to a circuit that receives certain sig-

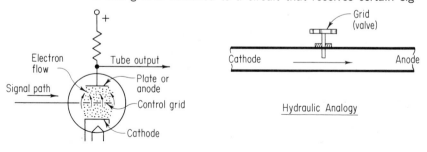

Fig. 3·19   Action of electron flow in vacuum tube.

nals, the grid potential may rise and fall, causing alternating changes in the flow of electrons toward the positively charged plate. Because of this throttling, or valvular, action, the electron tube is sometimes referred to as a valve.

3·16 Tube elements

The triode tube (whose action is described in Fig. 3·19) is so named because it has three electrodes—a plate, a cathode, and a grid. It also has a heater symbol, but the heater in this case is not called an electrode. In this instance, the cathode is the indirectly heated type—it consists of a thin metal sleeve which surrounds a heater of a tungsten-alloy wire filament. Sometimes the heater symbols are shown in a separate heater-circuit diagram and thus do not appear on the tube symbol.

(a)

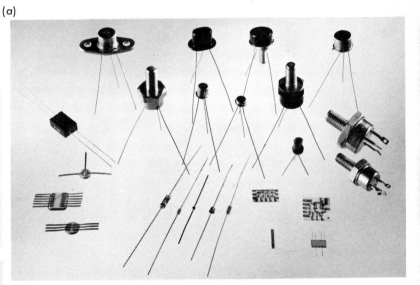

(b)

Fig. 3·20   Electron tubes and solid-state (semiconductor) devices. (a) Transistor, diode, and integrated semiconductor circuits. In the upper part of the photograph, the devices having three leads are transistors. The smaller devices in the lower part are diodes, and most of the flat-shaped objects are integrated semiconductor circuits. (General Electric Co.) (b) Electron tubes. Model 12HG7 is a frame-grid-type sharp-cutoff pentode in a T9 envelope. Model 7586 is a medium-mu Nuvistor triode in ceramic-and-metal construction. (RCA, Electronic Components and Devices Division.)

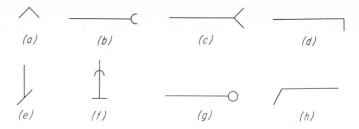

Fig. 3·21 Symbols for tube electrodes. (a) Directly heated cathode (or heater). (b) Photocathode. (c) Deflecting electrode. (d) Excitor. (e) Target or X-ray anode. (f) Dynode. (g) Cold cathode. (h) Ignitor, or starter.

Also, a tube may be directly heated, in which case the heater element itself is the cathode. If it is the heater symbol shown in Fig. 3·21a, it is drawn without the cathode symbol. The included angle in this symbol is 90°; the angle of the X-ray, or target, electrode is 45°. The electrodes shown in Fig. 3·21 are from the American National Standard Y32.2 and are reproduced in the back of the book along with many examples of different kinds of electron tubes. The circular tube symbol is

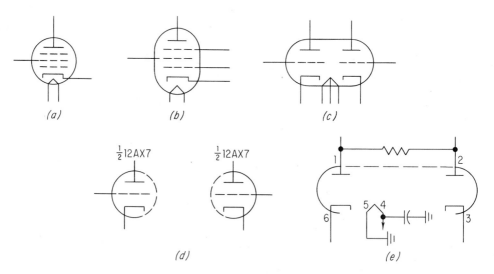

Fig. 3·22 Tube-envelope symbols. (a) Pentode in circular envelope. (b) Pentode in elongated envelope. (c) Twin triode in elongated envelope. (d) Multiunit triode in split envelope. (e) Multiunit diode in split envelope.

usually drawn a little larger than the transistor symbol. In many commercial drawings, it is made about 1 to 1½ in. in diameter. Sometimes the circle is made with a heavy line to make the tube (or transistor) stand out.

3·17 Tube envelopes

In most cases, a tube envelope is drawn as a circle. There are numerous instances in which this is not possible or desirable; in such cases, it is drawn as an elongated or split envelope. Figure 3·22 shows examples of elongated and split envelopes. The pentode can be drawn in either the circular or the oblong envelope, but it is less crowded in the oblong. Figure 3·22d and e shows different methods of splitting tube symbols. The tube represented by Fig. 3·22c and d is a twin triode.

3·18 Tube terminal connections

It is sometimes desirable to show the electrical relationship of elements of tubes to other parts of a circuit by means of numbers. This is accomplished by placing the number of a lead just outside the enclosure symbol, as shown in Fig. 3·22e.

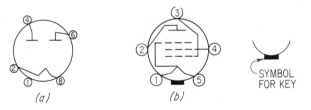

(a) (b) SYMBOL FOR KEY

Fig. 3·23 Base terminal markings. (a) Small pins. (b) Large pins.

Sometimes it is necessary to show the pin markings on tube symbols. Two sizes of circles—one for large pins, one for small pins—are authorized. Pin designations are numbered clockwise from the key. They may be obtained for nearly all tubes by referring to a tube manufacturer's manual or catalogue. Note that pins 3, 5, and 7 of the octal-tube base in Fig. 3·23a have not been used and hence are not numbered.

3·19 Picture tubes

(Cathode-ray, kinescope, or video tubes)

There are two types of video or CRT tubes: electrostatic and

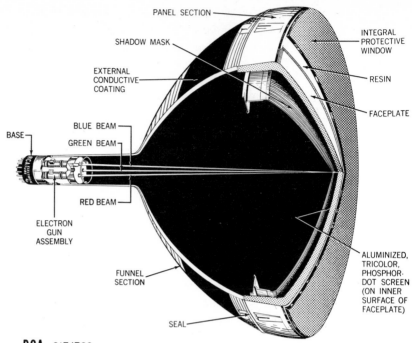

PANEL SECTION

INTEGRAL PROTECTIVE WINDOW

SHADOW MASK

RESIN

EXTERNAL CONDUCTIVE COATING

FACEPLATE

BASE

BLUE BEAM

GREEN BEAM

RED BEAM

ELECTRON GUN ASSEMBLY

FUNNEL SECTION

SEAL

ALUMINIZED, TRICOLOR, PHOSPHOR-DOT SCREEN (ON INNER SURFACE OF FACEPLATE)

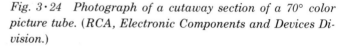

RCA - 21FJP22

Fig. 3·24 Photograph of a cutaway section of a 70° color picture tube. (RCA, Electronic Components and Devices Division.)

electromagnetic. Both types have three main parts to their system—an electron gun, deflecting devices, and a fluorescent screen. The electrostatic system is used primarily in oscilloscopes, while the electromagnetic system is used in most television tubes and for radar. Picture tubes are generally portrayed symbolically, as shown in Fig. 3·25. In Fig. 3·25*a*, the deflecting electrodes are shown, while they are omitted in Fig. 3·25*b*. In this figure, however, dashed lines signifying conductive coatings on the glass are shown. Accelerating voltage is applied to the inside coating through a high-voltage connector on the outside of the tube. The outside coating, grounded to prevent radiation, forms a capacitance with the inside coating to filter high-voltage pulses applied to the inside of the tube. In this symbol, the deflecting electrodes are not shown; presumably they are shown elsewhere, separately—a not unusual practice. Figures 3 · 25*c* and *d* show another way of drawing picture-tube symbols. These follow the standard

(a)

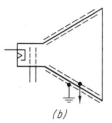

(b)

(c)

(d)

Fig. 3·25 Picture-tube symbols. (a) and (b) Electromagnetic. (c) Electrostatic. (d) Electromagnetic.

more closely than do Figs 3 · 25a and b, but most diagrams indicate the tube shape as it is shown in Fig. 3 · 25a and b.

3·20 Gas-filled tubes

The action of gas-filled electron tubes is considerably different from that of vacuum tubes. In the gas-filled type, many more electrons reach the anode than leave the cathode. This so-called *gas amplification* occurs because atoms of the gas become ionized. In one much-used tube, called the thyratron, a triggering or firing action takes place, following which the tube acts as a rectifier. A large dot in a tube symbol indicates that it is gas-filled.

3·21 Nomenclature

Tubes, and to a lesser extent semiconductors, can be classified according to the number of electrodes as follows:

Diode—two electrodes	Pentode—five electrodes
Triode—three electrodes	Hexode—six electrodes
Tetrode—four electrodes	Heptode (pentagrid)—seven electrodes

3·22 The "istor" family

The following devices are either experimental or proved special-purpose semiconductors. Descriptions of these devices and the names of the manufacturers are given in the May, 1964 (vol. 7, no. 3), issue of *Electronics Illustrated.*

Binistor	Gyristor	Sensistor
Bolomistor	Krypistor	Squaristor
Calorimistor	Madistor	Stabistor
Chargistor	Memistor	Strainistor
Chronistor	Mistor	Surgistor
Cryosistor	Negistor	Thermistor
Dynistor	Nesistor	Thyristor
Ferristor	Neuristor	Trigistor
Field-effect varistor	Nuvistor	Trinistor
Filmistor	Oscillistor	Twistor
Frigistor	Persistor	Vamistor
Gaussistor	Raysistor	Varistor

CAPACITOR　　　PHONE JACK　　　SWITCH, ROTARY

CAPACITOR,
ELECTROLYTIC　　LAMP BULB, NEON　TRANSFORMER,
IRON CORE

CAPACITOR,
VARIABLE　　　LAMP BULB,
ILLUMINATING　POWER
TRANSFORMER

CRYSTAL,
PIEZOELECTRIC　　METER　　TRANSFORMER, ADJ.
POWDERED IRON CORE

DIODE　　　RESISTOR　　TRANSFORMER,
ADJUSTABLE CORE

INDUCTOR　　RESISTOR, ADJ.
(POTENTIOMETER)　　TRANSISTOR

JACK, PHONO　　SWITCH, SPST
OR DPDT　　TUBE

*Fig. 3·26 Drawings of some commonly used components.
(Artwork courtesy Heath Company.)*

SUMMARY   Nearly every electrical and electronics device has a standard symbol which can be used to represent the device in a diagrammatic drawing of a circuit. These symbols are drawn with medium-weight lines but can be made heavier, if it is necessary to highlight the symbol. Transistor symbols may or may not include the circle envelope, while tube symbols must have the envelope symbol. Theoretically, there are no particular sizes to which symbols should be drawn. Practically, though, there are minimal and maximal sizes for any symbol or set of symbols, and within a drawing, symbols should be drawn in correct relative sizes. Usually a device is represented by a symbol of one size throughout a drawing. Not more than two sizes of a symbol are recommended for a drawing. There are certain basic electrical and electronics devices with which a reader or drawer of electrical drawings should become familiar. Orientation of a symbol depends upon the direction of the conductor path along which it is placed. Many transistors perform the same function as their tube counterparts.

QUESTIONS

3·1   What authority (publication) or authorities would be good source material for the selection of symbols to be used in an electrical drawing?

3·2   How would you go about determining the size to make a symbol for a circuit breaker in an electrical drawing?

3·3   What determines the position (orientation) of a symbol in a drawing?

3·4   When would you put the polarity markings of a battery on its symbol in a drawing?

3·5   What would be a typical designation that would be placed in the rectangular resistor symbol? In a circular relay-coil symbol?

3·6   How many "points" would you make if drawing the zigzag resistor symbol? How many turns would you make on the helical inductor symbol? On the other inductor symbol?

3·7   Where more than one symbol is approved for an element (or component), how would you go about determining which symbol to use on an electrical drawing?

3·8   Assume that you are required to draw the line representing a signal path of a circuit approximately $\frac{1}{64}$ in. wide. How wide a line would you use to draw the symbols in that circuit drawing? Why?

3·9   What determines whether a ground symbol or chassis symbol should be used in an electrical drawing?

3·10   A statement in ANSI Y14.15 indicates that not more than two sizes can be used for any one symbol in the same drawing. Describe a situation that, in your opinion, would require two different sizes of the same symbol in a drawing.

3·11   In addition to the fact that it is obsolete and not standard, the old capacitor symbol of two parallel lines has another disadvantage. What is it?

3·12   Name five different kinds of switches or devices that can be used for switches in electric circuits.

3·13   How would you show a device in a circuit drawing if no known symbol for that device exists?

3·14   What does the dot signify when placed within the tube circle symbol?

3·15   What is the difference between a pentagrid and a pentode tube?

3·16   What is the difference between a diode and a triode tube? Between a tetrode and a diode?

3·17   Name some typical types of tubes that would lend themselves to representation by the elongated envelope symbol when graphically portrayed.

3·18   What is the drawing size of the circular part of a transistor symbol as compared to the drawing size of an electron-tube circle symbol?

3·19   Symbolically, how is a polarized capacitor shown in a circuit path?

3·20   What does the bar in the diode symbol represent?

3·21   Which lead of a PNP transistor represents the output?

PROBLEMS   The problems below will require that the student make use of the American Standard or the Military Standard which indicates the correct symbols. The symbols shown in Appendix C are taken from several American Standards. It would also be advantageous for the student to look at some of the schematic drawings shown at various places in the text, especially in Chap. 6, before beginning to solve the problems below.

Cross-section paper with four or five divisions to the inch may be helpful, especially if the symbols are to be sketched freehand. It is not an absolute requirement, however. Any type of detail or tracing paper or film will be adequate. Exercises can be drawn on 8½ × 11 paper, except where otherwise indicated.

3·1   Draw lightly three horizontal lines 9 in. long and 3 in. apart. At 2-in. intervals (starting ½ in. from the left end), draw the symbols for the following elements:

a. Capacitor (fixed)   b. Resistor (general)
c. Battery (multicell)   d. Speaker
e. Voltmeter   f. Transformer (general)
g. Relay coil   h. Fuze
i. Contact (normally   j. Motor
    open)
k. Transformer (general)   l. Switch (single pole)
m. Current transformer   n. Inductor (ceramic
o. Antenna   core)

Add the name below each symbol.

3·2   Follow the instructions given in the preceding problem, but show the following symbols:

a. Shielded capacitor   b. Chassis connection
c. Transformer with core   d. PNP transistor
e. Contact (normally   f. Inductor with two taps
    closed)
g. Polarized relay   h. Breakdown diode
i. Adjustable capacitor   j. Push-button switch
k. Variable resistor   l. Switch, double-throw
m. Tetrode transistor   n. Split-stator
o. Inductor with ceramic   capacitor
    core

3·3   Follow the instructions given in Prob. 3·1, but show the following symbols:

a. Capacitive diode   b. NPN transistor

c. N-type unijunction transistor   d. Delay function

e. Pool-type vapor rectifier   f. Incandescent lamp

g. Field-effect transistor, N-type base   h. Thermistor

i. Pentode tube   j. Semiconductor thermocouple, current measuring

k. Cold-cathode ac glow lamp   l. Thyratron tube (gas-filled triode)

m. PNPN transistor   n. Unijunction transistor, P-type base

o. PNN transistor, with ohmic connection

3·4   Follow the instructions given in Prob. 3·1, but show the following symbols:

a. Dipole antenna   b. Schmitt trigger

c. Separable connectors   d. Slow-operate relay

e. Breakdown diode, uni-directional   f. Capacitor with split stator

g. Pickup head, recording   h. Switching function, transfer

i. Thermistor, general   j. Adjustable resistor

k. Field-effect transistor, N-type base   l. Twin triode tube

m. Four-conductor shielded cable   n. Tunnel diode

3·5   Sketch or draw the following tube symbols, placing the name under each tube. Use indirectly heated cathodes.

a. Diode   b. Tetrode

c. Thyratron (gas-filled triode)   d. Twin triode with common cathode

e. Gas-filled ignitron (with exciter and control grid)   f. Triode

g. Pentode   h. Twin diode

i. Pentagrid or heptode

3·6 Redraw the symbols shown in Fig. 3·27 at about twice the size they appear in the book. (Use one 11 × 17 or 12 × 18 sheet, or two 8½ × 11 sheets.) Then

Fig. 3·27 (Prob. 3·6) Symbol identification exercise.

SYMBOL   IDENTIFICATION

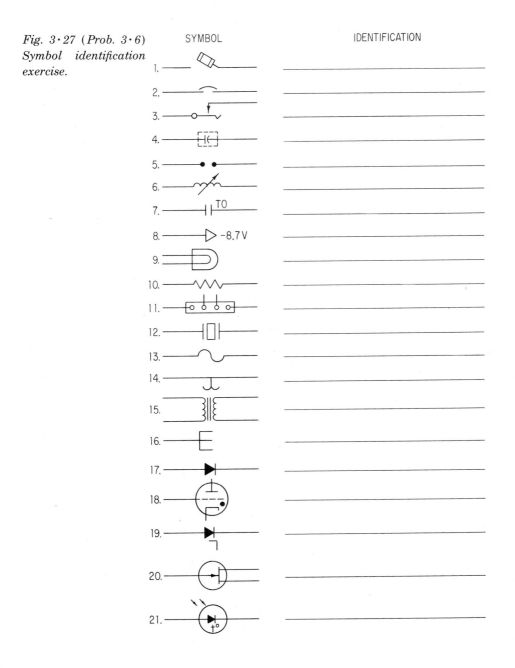

identify each device symbolized at the right. (*Suggestion:* Draw a horizontal line to the right of each symbol, as shown in Fig. 3·27, and letter the name along each line.) Identification, in many cases, should include more than just the name. If it is a transformer, for example, what *kind* of transformer it is; if a switch, what type it is and what its operating condition is.

3·7 Draw the circuit shown incompletely in Fig. 3·28, and put the symbols in the places shown, correctly positioned. Where figures 1, 2, and 3 appear, draw resistors. At 4, draw a capacitor (fixed). At 7, draw a variable capacitor. At 5, draw a PNP transistor, with emitter at *e*, and so forth. At 8, draw an inductor. Place identifying lettering at terminals as follows: 11, *J*10; 15, *GRD*; 16, +5; 12, *J*15; 13, *J*20; 17, *J*21; and 14, −5. All lines in the completed diagram should be the same weight. If necessary, go over lines that are too light. Use 8½ × 11 paper.

3·8 Complete the circuit shown in Fig. 3·29 by adding the symbols listed below at the places where the numbers appear.

Oil circuit breakers (with male connectors): 1, 2

Connection: 3

Arrester: 4

Ground: 5

Step-down transformer: 6

Rectifier, SCR: 7

Circuit breaker general: 8

Bypass switch, general: 10, 11

Potential transformer: 9, 12

The completed drawing should have all lines the same weight, with the exception of the 3/0 feeder wire which may be thicker. Use 8½ × 11 paper.

3·9 Draw the circuit shown in Fig. 3·30 about twice as large as it appears there. Then add the symbols as indicated by their standard letter-number designations. (These can be found in Mil Std 16B, shown in Appendix A and elsewhere in the book.) The transistor is of the NPN type, with base, emitter, and collector leads identified. Use the terminal symbol at the input and output. *C*1, *C*3, *C*4, and *C*6 are variable. *L*1 and

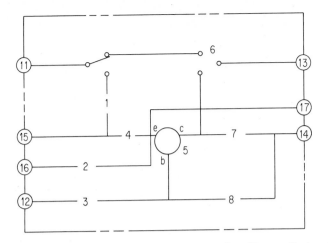

Fig. 3·28 (Prob. 3·7) Symbol exercise. Transmit-standby Apollo color camera. (No-dot system is used.)

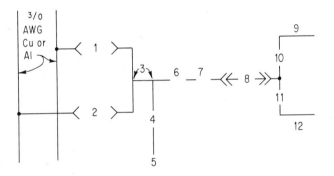

Fig. 3·29 (Prob. 3·8) Symbol exercise. BART substation.

Fig. 3·30 (Prob. 3·9) Symbol exercise. Power-amplifier circuit.

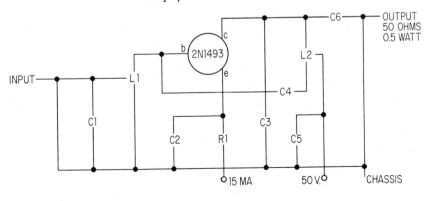

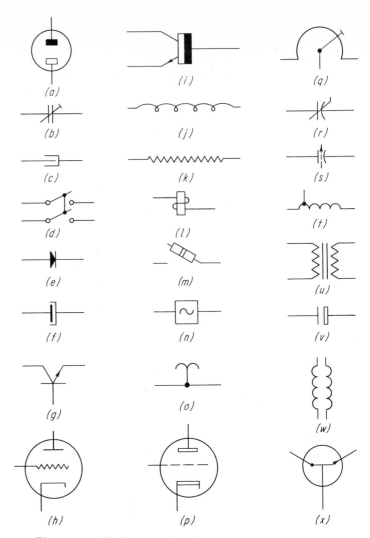

Fig. 3·31   (Prob. 3·10) Symbol exercise. Incorrect symbols.

L2 have fixed taps. The lines on the completed drawing should have the same weight. Use 8½ × 11 paper.

3·10   Each symbol in Fig. 3·31 is incorrect or not American National Standard. First, identify the symbol that is being depicted. Then draw the correct symbol and label it, using the same general layout as in Fig. 3·31. Use 8½ × 11 or 11 × 17 paper.

3·11   Identify some symbols assigned by your instructor. Use the format assigned by him.

Chapter 4
Production drawings

Before an electronics/electrical package is built, instructions must be given to the production workers who are going to have to build or assemble the unit. Drawings convey these instructions very well, although they sometimes must be accompanied by other information. There are several different kinds of drawings used for this purpose. Correct or recommended practice for most of these drawings is set down in ANSI Y14.15 and Mil Std. D-16415. We shall show and explain most of the different types.

Connection drawings

4·1 Connection, or wiring, diagrams

In order to connect the various component devices (resistors, transistors, etc.) in an instrument such as an amplifier, it is the usual custom to follow a *connection diagram*. The older name for this type of drawing is *wiring diagram,* and because it is so widely used, we shall use these two terms interchangeably.

In a manufacturing company these drawings are used by the time-and-motion, quality-control, and estimating sections, as well as by individual workers. Wiring diagrams are also used in the maintenance area by persons doing checking, trouble-shooting, and modification. Sometimes they are used with schematic diagrams (explained in Chap. 6) to provide additional information.

In mass production, a connection diagram may be used at first, then put away after the worker has memorized the steps and connections. In some instances, companies with active methods sections are able to produce without such a drawing.

4·2 Types of connection diagrams

These diagrams may be classified according to their general layout as follows:

1 Point to point
2 Highway, or trunkline
3 Baseline, or airline

There are also other ways of labeling wiring diagrams, as the reader will discover. Such titles might include *cabling, harness,* and *interconnection* diagrams.

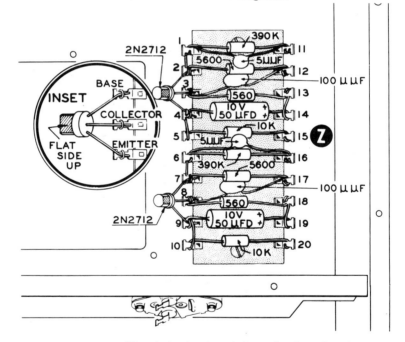

Fig. 4·1 A pictorial production drawing for wiring an electronic assembly. (Heath Company.)

4·3 Point-to-point diagrams

This is the oldest type of wiring drawing. Figure 4·1 shows a pictorial form of the diagram in which each part and terminal

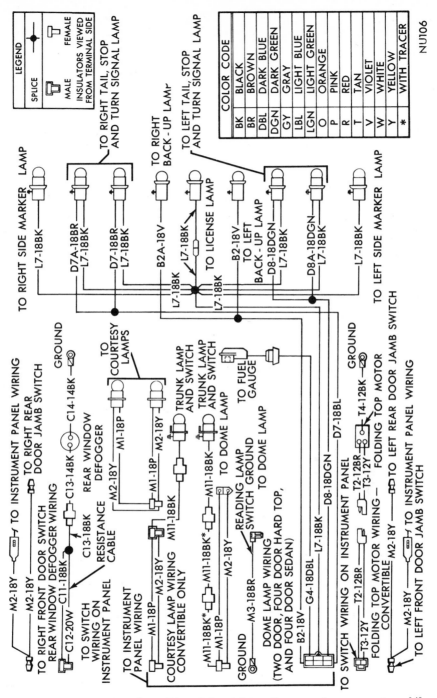

Fig. 4·2 *Pictorial wiring diagram for an automobile.* (*Chrysler.*)

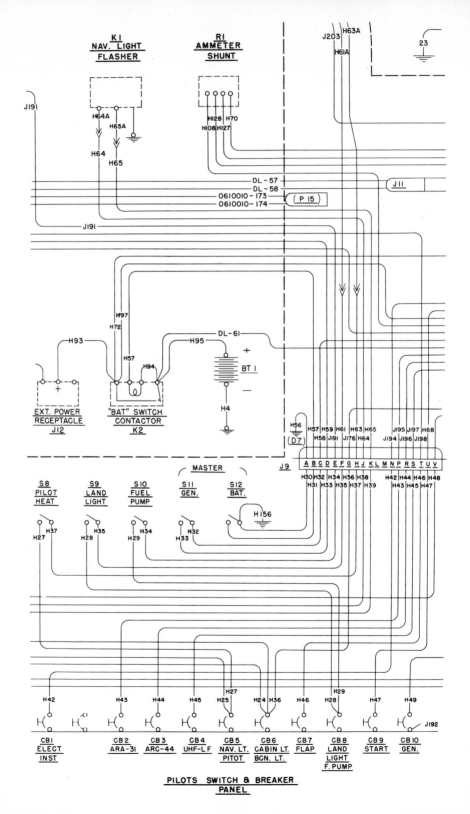

PILOTS SWITCH & BREAKER
PANEL

72

are drawn about as they appear to the viewer. This type of drawing has met with great success in the do-it-yourself kit industry and other places.

Figure 4·2 is another type of pictorial point-to-point diagram, but it is not as close to what the actual object looks like as is the first drawing. Most of the symbols are pictorial, but their relationship to each other is not entirely accurate. Furthermore, the lines representing the wires (conductors) are drawn either horizontally or vertically, not as they actually are placed in the chassis or frame of the automobile. A distinctive feature of this drawing is the color code and identification system required when there are as many wires as are presently utilized in motor vehicles. The wire in the upper right corner, for example, is *L7-18 BK*. This means that this is conductor *L7*, which is of size 18 wire (see p. 360) and is black in color.

Fig. 4·3 Connection (wiring) diagram for a small aircraft. (Cessna.)

In the two drawings discussed so far, the component parts have been drawn pictorially. In most wiring diagrams, these parts are shown by means of symbols or by simple geometrical figures such as squares, rectangles, and circles. This will be the case with the connection diagrams that follow, including the aircraft wiring diagram of Fig. 4·3.

Figure 4·3 shows only a small part of the original drawing, but it contains enough details to give a fair picture of what such a drawing looks like. Note the switch symbols *S8*, *S9*, etc., the battery symbol, *BT*1, and the circuit-breaker symbols, *CB*1, *CB*2, etc. Note, also, the uniform spacing of wire conductors, drawn about ¼ in. apart, with their rounded "corners." Actually, most of the wires are routed in bundles (harnesses), and the physical routing of wires may or may not bear close resemblance to the way they are drawn here. Item *J*9, with letters A through V, is a receptacle into which another plug receptacle, *D*7, is plugged. The letters represent the various pins in the plug. The reader will have a difficult time tracing the paths of many of the wires in this drawing because only part of the drawing is here. He can trace *H*27 from switch *S*8 to *CB*5 and line *H*37 from *S*8 to *J*9, however. Unlike the first figure, Figs. 4·2 and 4·3 do not utilize extra-heavy lines to represent wires. All line work, including symbols, is uniform—a practice widely followed in industry.

4·4 Interconnection (external wiring) diagrams

Continuing with examples of point-to-point wiring diagrams

is Fig. 4·4, which consists of conductors (wire or cables) connecting items of equipment or unit assemblies. Internal connections are customarily omitted in such a drawing. Note

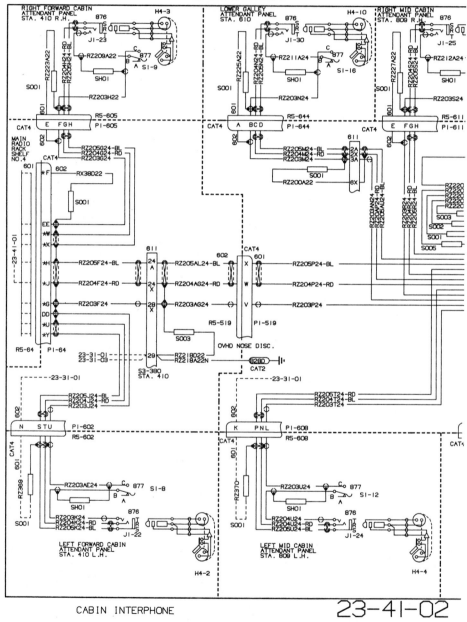

CABIN INTERPHONE

23-41-02

that the conductor lines are either horizontal or vertical and that when they change directions, the corners are "sharp," a practice which is as popular as the rounded-arc pattern shown in Fig. 4·3. Another good example of interconnection diagrams is Fig. 1·1.

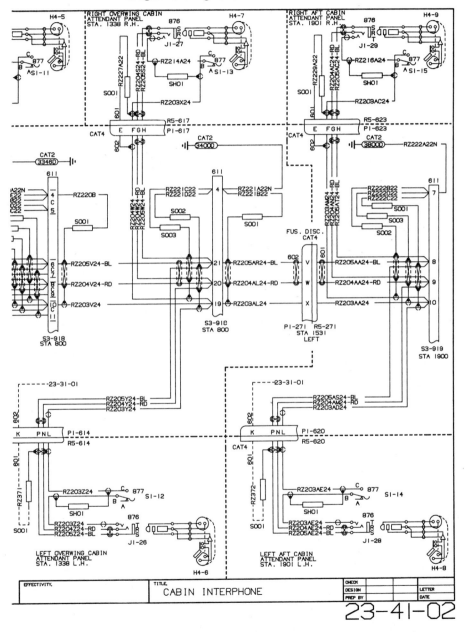

Fig. 4·4   Interconnection diagram. DC-10 cabin interconnection system. (McDonnell-Douglas.)

4·5 Highway, or trunkline, type of wiring diagram

Practically any electrical assembly or system can be shown wired or connected graphically by the point-to-point method. But the same assembly or system can also be connected graphically by other methods. Sometimes the point-to-point diagram gives the clearest picture and fastest reading, yet at other times, one of the other methods might give the best picture and provide easier and quicker reading. One of these other methods is the highway type of connection diagram. This type differs from the point-to-point style in that *conductors* (*conductor paths* might be better) are merged into long lines called *highways* (or *trunklines*) instead of being drawn as separate, complete lines from terminal to terminal. The short lines leading from the terminals to the highways are called *feed,* or *feeder, lines,* and must have some sort of identification near the highway so that the conductor path can be followed by the reader.

One or more highways may be shown, depending upon the wire routing necessitated by the physical (or graphical) arrangement of components. These trunklines do not have to conform to actual bundles or harnesses, if such are used. Figure 4·5 shows the rear view of a control panel of a copy-

Fig. 4·5 A typical highway type of connection diagram.

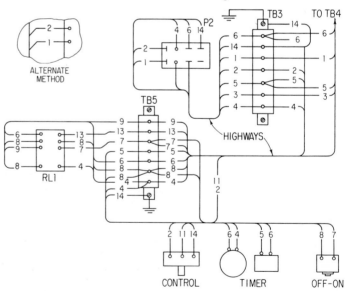

reproducing machine. In this figure, the highways are located so that the lengths of the feed lines are short and so that there is a minimum of crossing of highways and feeders. (Highways may cross, and feeders may cross each other and highways, but the number of crossings should be kept to a minimum.) Each conductor is given a number, and that number appears periodically on the drawing so that the reader can follow the conductor path. Line 1, for instance, can be traced from P2 to TB3 and on to TB4, which does not appear on the drawing. Line 2 can be followed from P2 to the control terminal. It must be remembered that there is a separate wire for line 1 and another wire for line 2 from P2 to TB3. The highway is just an imaginary thing. The direction a conductor path takes when it is in the highway is indicated by the direction of the quarter-circle arc where the feed line enters the highway. For example, when leaving the control terminal, line 11 goes to the right and line 14 goes to the left. Some companies use short 45° lines instead of circle arcs, as shown in the inset of Fig. 4·5.

A different method of identifying circuit paths is in wide use. This ANSI-recommended method suggests that each feed line have this designation near the point where it joins the trunkline: (1) component destination, (2) terminal or component part, (3) wire size or type (if necessary), and (4) wire color. Referring to Fig. 4·6, the top feeder line at terminal board 12 carries the following identification: TB14/4-B2. Broken down, this would be: TB14—destination; 4—terminal at destination; B—size of wire (say, No. 22 hookup wire by a prearranged code); and 2—color (red, by Mil Std 122 color code shown on the drawing and also Appendix B). Going to TB14 and terminal 4, we find the other end of this connector, and here it has the reverse identification, TB12/1-B2.

The choice of location and number of highways may depend on several factors. Separate highways for different cable groupings may be desirable. If a group of wires needs to be segregated or shielded, a separate highway for that group may be desirable. Such a highway might be drawn close to, and parallel to, another highway. Occasionally situations may arise where certain wires, called *critical wires,* should be drawn from point to point and not merged into a highway. Special identification for these wires may be necessary.

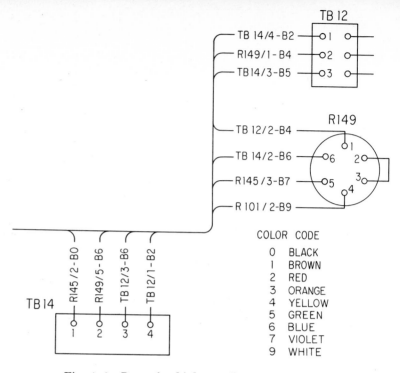

TB 12

TB 14/4 -B2 — 1
R149/1 -B4 — 2
TB14/3-B5 — 3

R149

TB 12/2-B4 —
TB 14/2 -B6 —6   2
R145 /3-B7 —5   3   4
R 101 / 2-B9 —

TB 14

R145/2-B0
R149/5-B6
TB 12/3-B6
TB 12/1-B2

1   2   3   4

COLOR CODE

0   BLACK
1   BROWN
2   RED
3   ORANGE
4   YELLOW
5   GREEN
6   BLUE
7   VIOLET
9   WHITE

Fig. 4·6 Part of a highway diagram using American National identification system.

The highway type of connection diagram is particularly suited for showing the wiring of large panels where there are many terminals and conductors. Figures 4·5 and 4·6 are both panel-wiring diagrams. A panel-wiring diagram generally shows the back, or wiring, side of the panel. Terminal strips, switches, breakers, etc., are usually shown in exact or approximate position with regard to each other. Sometimes the panel is drawn as a scale drawing showing all terminals, lights, lettering, and switches, and the wiring is shown diagrammatically elsewhere on the drawing sheet.

4·6 Airline, or baseline, diagrams

These wiring drawings, with a rather misleading name, are similar in some ways to highway diagrams. The *airline,* or *baseline,* as it is sometimes called, is an imaginary, usually horizontal or vertical, line conveniently located so that short feed lines may be drawn from component terminals to it.

The heavy dark line in Fig. 4·7 is the airline in this drawing. The feed lines are drawn at right angles to it, and there are no curves where they join the airline, as there are in highway diagrams. As in the case of highway diagrams, good identification of feed lines is necessary. In this figure, a simplified version of the American National system is used. Each feeder has the number of the destination component part and the color of the wire. The drawing is read as follows: Take the conductor, which is identified as BK-R 13, leading from part 15. Now go to component 13 and look for the black-red feeder. It is quickly found with the identification BK-R 15. So we have an identification and reading procedure similar to that used in highway diagrams. The major difference between airlines and highways is that a highway must go from one component to the destination component, sometimes making several bends and turns to get there, but an airline may stop at any convenient place and another airline may be drawn near the destination component. In our example this does not happen, but in some airline drawings having many component parts there are many airlines spotted at different places close to groups of parts. In such cases, a feed line may enter one airline and come off another airline. This makes for slower reading and tracing of conductor paths, but for systems or packages containing hundreds of wires and terminals, the airline type of diagram may be less cluttered

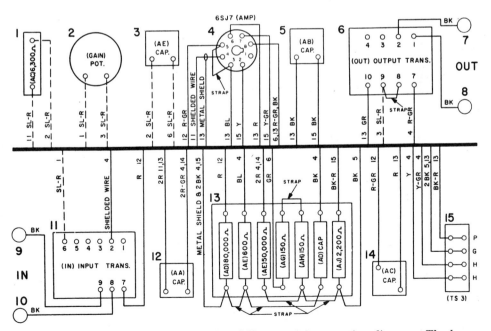

Fig. 4·7 An airline type of connection diagram. The heavy line is the airline. (It is often not made heavy.)

and confusing than point-to-point and highway drawings are.

Identification is important in two respects. First, the component parts themselves must be labeled with large block numbers in or near their upper left corners. And second, proper letters should be used for colors where a numbered color code is not used. One-letter designations are adequate for some colors such as yellow and red, but two letters are necessary for others such as black (BK), blue (BL), brown (BR), green (GN), gray (GY), and slate (SL). Figure 4·7 follows this concept fairly well except that GR is used for green, which is a bit confusing, although SL (slate) is used for gray. (ANSI Z32.13, "Abbreviations for Drawings," specifies G for green and GY for gray.)

Frequently, so many components are present in a large system that if they were shown as one column or strip, the airline diagram would be too long and narrow for practical purposes. To avoid this, the diagram may have to be laid out in several columns side by side. In some cases, a feed line may represent more than one wire, as witness the feeder coming from terminal 8, component 4. Some of the feeders in the left part of the diagram are shown with dashes. They happen to be part of a feedback circuit which is being emphasized. This is not very common practice. In both highway and airline diagrams, straps or "pigtail" leads are often shown. These go from one terminal of a component to another and are often identified as *straps* or *PTs*.

4·7 Other examples of connection diagrams

Frequently, it is desirable to show the wiring diagram and the schematic (elementary) diagrams on the same drawing. Figure 4·8 shows these two diagrams for a motor controller. In this panel-wiring drawing, two different line widths are used for conductors. Narrow lines, say ¹⁄₁₀₀ in., are used for the control, or pilot, section of the circuit; and wide lines, say ¹⁄₅₀ in., represent the line-voltage part.

A *cabling diagram* of an industrial TV installation is shown in Fig. 4·9. Actually, it is just a variation of the interconnection diagram. The corners of the connecting conductors have been rounded to simulate cable. If drawn to scale, this could also be called a *location diagram*. Any wiring diagram that is drawn to scale, or approximately so, may be classified as

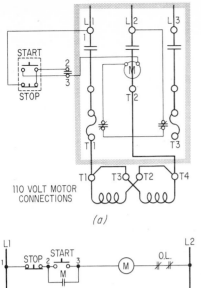

110 VOLT MOTOR
CONNECTIONS

START

STOP

(a)

Fig. 4·8 Connection (a) and elementary (b) diagrams of a single-phase starter. (From Thomas E. French and Carl L. Svensen, "Mechanical Drawing," 6th ed., McGraw-Hill Book Company, New York, 1957.)

L1   STOP START   O.L.   L2
1   2   3   M

M

(b)

Fig. 4·9 A cabling diagram of an industrial television installation. (Allen B. Du-Mont Laboratories, Inc.)

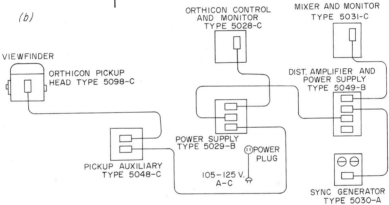

ORTHICON CONTROL
AND MONITOR
TYPE 5028-C

MIXER AND MONITOR
TYPE 5031-C

VIEWFINDER

ORTHICON PICKUP
HEAD TYPE 5098-C

DIST. AMPLIFIER AND
POWER SUPPLY
TYPE 5049-B

POWER SUPPLY
TYPE 5029-B

POWER
PLUG

PICKUP AUXILIARY
TYPE 5048-C

105-125 V.
A-C

SYNC GENERATOR
TYPE 5030-A

a *location diagram.* This term is not used as extensively in the United States as it is in the British Commonwealth. Additional information, such as sizes and types of cables and types of plug connectors, is often shown in cabling diagrams.

For very complex installations, it may be necessary to compile *wiring lists* or *to-and-from* diagrams in addition to, or in place of, connection diagrams. A typical heading for such a list might be:

ITEM NO.	SYMBOL	COLOR	SIZE	FROM		TO	
				Component	Terminal	Component	Terminal

Additional information, such as routing (conduits, ducts, etc.), function, and Military type number, may be shown.

4·8 Line spacing and arrangement

The reader may or may not have noticed the neat, uniform line spacing in some of the examples—particularly the point-to-point diagrams. This spacing and layout were not achieved accidentally but rather by careful planning and good use of drafting aids. Two especially helpful aids are (1) the preliminary freehand sketch (or sketches), and (2) a commercially produced or office-made undergrid.

When making the original sketch, it is advisable to consider laying out the components so that the simplest, neatest wiring diagram will result. This may include changing the location of the components with regard to each other (if such juggling is permitted) and changing the order of the terminals. Rather elementary graphic examples are shown in Fig. 4·10. The shorter and more direct the lines are, the quicker and easier it is for the reader to trace the paths. Keeping the crossings to a minimum also facilitates reading.

Spacing of parallel lines is most often at ¼ in. This permits use of a prepared undersheet having ¼-in. grids, which, if a transparent drawing medium is used, greatly facilitates the making of the final drawing. For various reasons (lettering requirements, reduction of drawing) other spacings may be required.

4·9 Wiring harness or local cabling

A large portion of the expense in producing complex electronics assemblies can be laid to the wiring. The wiring operation can be optimized (made simpler and cheaper) by study of the chassis and connection drawings, followed by the drawing of a harness diagram. This is generally done in a series of steps:

1. Study of the mechanical assembly, noting the general path which the harness should take and any obstacles which might cause difficulty
2. Making a drawing of the outline of the harness. This can best be done if a suitable full-scale chassis or assembly drawing exists. A sheet of tracing paper or other translucent drawing medium is then placed over the drawing, and the harness outline done on the top piece.

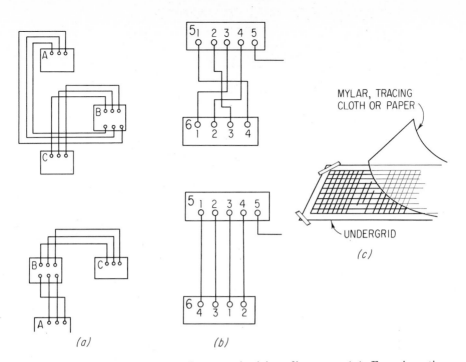

Fig. 4·10 *Layout of wiring diagrams. (a) Experimenting with location of components. (b) Changing terminal locations (lower part). (c) Using preprinted undergrid.*

3 Completion of the harness (sometimes called *local-cabling*) diagram itself, using the assembly drawing and the wiring drawing for the assembly. This may include "breakout" points where individual wires (or several wires) break out of (or enter) the harness and where nails or pegs will be driven into a board or jig on which the cabling will be assembled.

4 Proper identification of each wire on:
   a. The drawing itself
   b. A table or listing of wires by color, etc.

Figure 4·11 shows a typical local cabling, both by itself and as it later appears connected to the assembly. There is additional cabling and wiring on the same assembly.

Figure 4·12 shows how such a diagram may be begun, and Fig. 4·13 shows the final full-scale drawing of the harness with breakout points and lines showing where the wires are stripped. For example, the wires in the upper left corner extend beyond the line marked 104, 5 but are stripped back to this line, and the bare wire is cut so that ¾ in. remains above

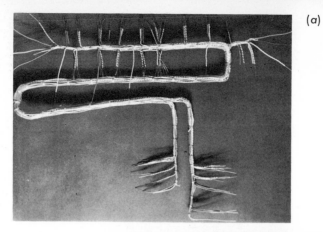

(a)

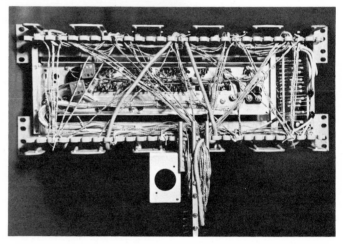

(b)

Fig. 4·11 Photograph of a local cabling harness. (a) Harness by itself. (b) Harness attached to chassis. (Western Electric Co.)

the line. The approximate thickness of the cabling is shown and is based on the number (and thickness) of wires that are in the harness at these places. If 26 wires are in the center

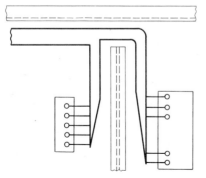

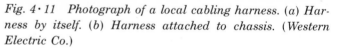

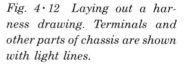

Fig. 4·12 Laying out a harness drawing. Terminals and other parts of chassis are shown with light lines.

Table 4·1   Harness-wire routing

color	start	break out at	finish		terminal data
BL	101		119		CAP(C34)
BL	120		114		TERMS
BL	111		121	103	1
BL	118		427	104	2
R	408		120	105	INDR
R	109		412		(L31)
R	428		114	106	CAP(C31)
R	115		111		REL(AL)
R	110		316		TERMS
R	304		128	107	77R
R	129		121	108	BOT 3, 4
BK	107		103	109	TOP 1, 2
BK	106	108,118		110	REL (DB)
		120,123	415		

NOTE: This tabulation is part of drawing shown in Fig. 4 · 13.

part, the approximate thickness there would be about $\sqrt{26}$ × ¹⁄₂₀ (thickness of insulated No. 22 wire), or about ¼ in. plus, which is only slightly less than what the finished harness measures at this point.

Table 4·1 shows the tabular form of wire identification and routing. It is often placed on the same sheet as the harness drawing. This system uses a sort of "station" method whereby leads numbered in the 100s are at one area of the harness and those numbered in the 300s and 400s are at other areas, or "stations," along the harness. Other numbering systems are sometimes used. Figure 4·13 is one of a series of drawings made for the assembly of the electrical equipment. Other drawings include a schematic diagram and connection diagrams.

Construction and assembly drawings

4·10 Sheet-metal layouts

In order that the frames, chassis, shields, and other such

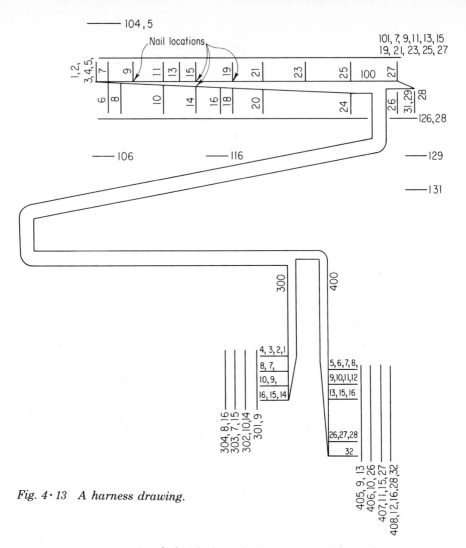

Fig. 4·13 A harness drawing.

parts of electrical or electronics assemblies can be correctly and economically manufactured, drawings must be made that will tell the builder exactly how the item is to be made. Figure 4·14 shows the pattern of a shield for some electronics equipment of a satellite. It is laid out flat; the lines on which it is to be folded are shown with double dashes. In addition to having complete dimensions, the drawing tells exactly what kind of material is to be used and how the material (an aluminum alloy) is to be finished. Although it does not tell how to fold the metal, a little study will reveal that this will make a boxlike enclosure if the sides are folded at 90° to each other. Chassis diagrams are treated in the same way, unless

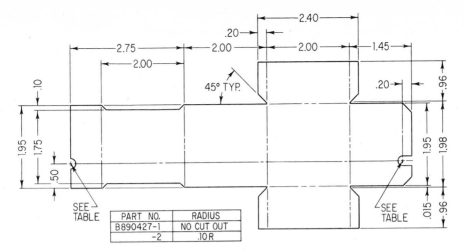

PART NO.	RADIUS
B890427-1	NO CUT OUT
-2	.10 R

Fig. 4·14   The pattern drawing of a shield for a communications satellite. (Bell Telephone Laboratories.)

they have too many dimensions to make the folding-out
graphical concept practical.

Figure 4·15 shows the same type of drawing for another
piece of satellite equipment, except that a different dimensioning system is used. Here, horizontal location dimensions are
given from the left edge because of the critical distances to
the square projections, which must fit into mating recesses

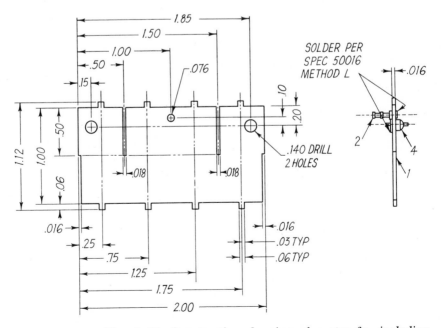

Fig. 4·15   Construction drawing of center fin, including
terminals. (Bell Telephone Laboratories.)

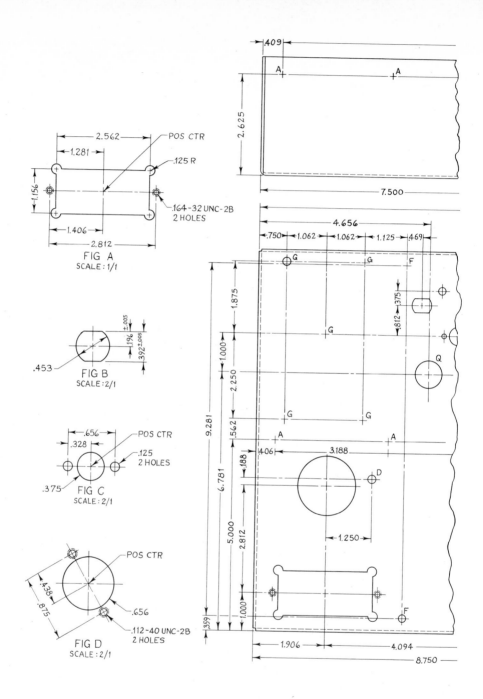

Fig. 4·16 Part of a drawing for the construction of a main chassis. (Western Electric Co.)

on another piece. This method of dimensioning to a well-defined datum plane (it should be on a finished surface) is more accurate than the other method of dimensioning from one feature to another, and so on.

4·11 Chassis manufacture

Figure 4·16 shows only part of a construction drawing for a chassis like that shown in Fig. 4·11, with the wiring harness. The complete drawing has four views, plus supplemental drawings, showing the exact shapes of some of the holes and other cutouts. Because of size limitations, we have shown only the left quarter of two adjacent views and some of the cutout details, which are placed around the edge of the sheet. The cutout dimensions are given in separate details because there is so much information on the main views that any more dimensions would make a cluttered drawing which would be difficult to read.

For complete manufacture of this chassis, a separate set of instructions, entitled Manufacturing Layout and Time Rate, is issued to the manufacturing section. Some of the 22 steps included in this Layout and Time Rate are shown in Table 4·2. Note that even the tools are specified.

This chassis drawing is a typical engineering drawing made to full scale ($1'' = 1''$). It can be used later on as a basis for an assembly drawing (for putting the rest of the chassis together) and for the harness, or local-cabling, drawing. The advantages of making it to full size are now fairly evident. However, chassis for some miniaturized packages cannot be drawn to full scale because they are too small. They must be drawn larger than actual size in order that details and dimensions can be appropriately shown.

4·12 Hole and terminal data

Chassis and wiring boards may be dimensioned according to standard drafting practice. There are several methods in use. Two systems that would be quite suitable for parts having many holes are shown in Fig. 4·17. These methods can be used for manual operation of standard drill presses or for numerical control (N/C) machine tools. Figure 4·17a illustrates the type of drawing that is ideal for a programmer to use. Various sizes of holes are indicated by different letters, and the holes within a letter group are numbered sequentially. The xy reference

Table 4·2 Manufacturing layout and time rate*

| RM 512261 (Stock No.) | Brass sheet, .063″ × 24″ × 84″ Grade A. (5) parts per sheet. (20) sheets per 100 parts. | |

operation description	*machines, tools, gages, test sets, etc.*

metal parts
1. Shear strips to 15.923 ±.010″ × 24″
   a. Gage part from front measuring scales on shear
      (backup).
   b. Place parts on pallet.
	Niagara shear (11005)
	Use front gaging.
	18″ vernier caliper

2. Trim parts to 22.674 ±.010″ × 15.923 a. Gage part from front measuring shear (backup).	Niagara shear (11005) Use front gaging. 24″ vernier caliper

3. Perforate blank complete except that (4) corner notches (cutouts) are not to be blanked at this time.
   (1) setup
   (2) handlings
	#5 Minster punch press (62221)
	P&D C-729515
	Bolster C-673918-9

4. Omitted	
5. Deburr part. a. Place part on pallet.	Bench Pneumatic sander
6. Omitted	

7. Tap (2) holes .164–32 for "B" cluster
   (1) Setup
   (2) Strokes
	Snow tapper (79003)
	(1) H.S.S. tap, .164–32
	(2) Flute plain pt.

*Five of twenty-two steps are described.

point is located close to a corner, in this case at the center of hole A-1. The programmer will be able to write the program (which will later—through tape—give instructions to the machine tool as to how to do the job) in a very efficient manner. Holes that are the same size will be drilled one after another. After these holes have been drilled, the drill bit will be changed (manually on some tools, automatically on others) and then the next series of holes will be made.

The table of Fig. 4·17a is so constructed that the *absolute* method of N/C can be easily performed. However, the other method, called *incremental*, may also be used by subtracting x and y values of one hole from the next. Figure 4·17b also

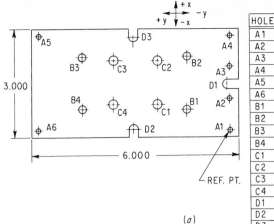

HOLE	X	Y	SIZE	TOL
A1	0.000	0.000	0.125 D	.003
A2	0.844	0.000		
A3	1.656	0.000		
A4	2.500	0.000		
A5	2.500	5.500		
A6	0.000	5.500		
B1	0.500	1.250	10-24	NC
B2	2.000	1.250		
B3	2.000	4.250		
B4	0.500	4.250		
C1	0.625	2.125	0.250 D	.001
C2	1.875	2.125		
C3	1.875	3.375		
C4	0.625	3.375		
D1	1.500	0.125	0.187 R	.002
D2	0.000	2.750		
D3	2.500	2.750		

(a)

HOLE	NO. REQD	SIZE
A	6	.125 D
B	4	10-24
C	4	.250 D
D	3	.187 R

(b)

Fig. 4·17 Positional dimensioning. (a) Coordinate system. (b) Baseline system.

lends itself to both absolute and incremental programming approaches. Actually, it is not necessary to draw the holes on this type of drawing. Some companies follow the practice of showing center lines only. Holes *B1* to *B4* are, of course, threaded. They are No. 10 taps of the coarse thread (NC or UNC) series having 24 threads to the inch.

4·13 Assembly drawings

Figure 4·18 shows the required components drawn in place on a board. Some objectives to be observed in locating these components are:

1 To fit the parts into the space available
2 To satisfy wiring requirements, such as accessibility for connections and short wiring paths
3 To achieve a pattern without crossings if printed circuitry is to be used
4 To satisfy electrical requirements such as shielding, built-in capacitances, etc.

Such a layout is made by (1) studying the original schematic, or elementary, diagram, to determine what components have common connections, where common lines (e.g., ground, B+) are, etc.; (2) drawing one or more point-to-point wiring diagrams freehand in order to get an optimal arrangement; (3) further refinement, if justified or necessary, which

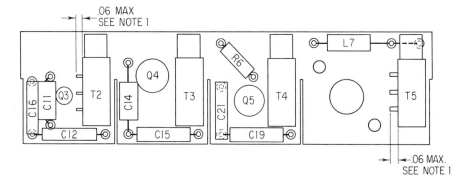

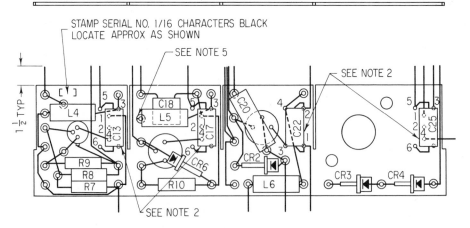

Fig. 4·18 An assembly drawing of components on a board. This is sometimes called a components location drawing. (Bell Telephone Laboratories.)

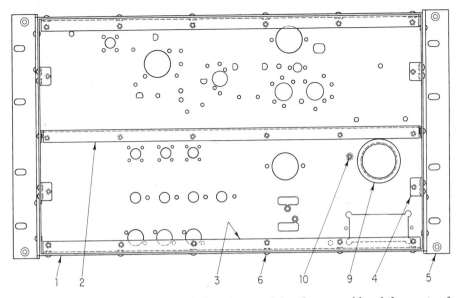

Fig. 4·19 A drawing used for the assembly of the parts of a complete chassis. (Western Electric Co.)

might include cutting out outlines of parts on heavy paper and shifting them into different arrangements; and (4) drawing the final assembly drawing, such as is shown, to scale. Each component is given a designation, usually letter-number, such as $R1$ (resistor No. 1) and $C5$ (capacitor No. 5).

This drawing can be used for two or more purposes: (1) to give a parts-location layout, along with a list of parts and catalogue numbers, to production and service personnel, and (2) to provide the location of holes for such a drawing as Fig. 4·17.

Another type of assembly drawing is that used for the assembly of a chassis itself. Figure 4·19 shows one of three views of such an assembly. The drawing also lists the various parts as follows: (1) chassis (also shown partially in Fig. 4·16), (2) panel, (3) and (4) angles, (5) brackets, (6) and (7) rivets, (8) grommet, (9) Penn fastener, and (10) self-clinching fastener. (Numbers 7 and 8 are shown on another view.)

As in the case of Fig. 4·16, this drawing is accompanied by a Manufacturing Layout and Time Rate sheet which lists each assembly step and the tools with which the operations

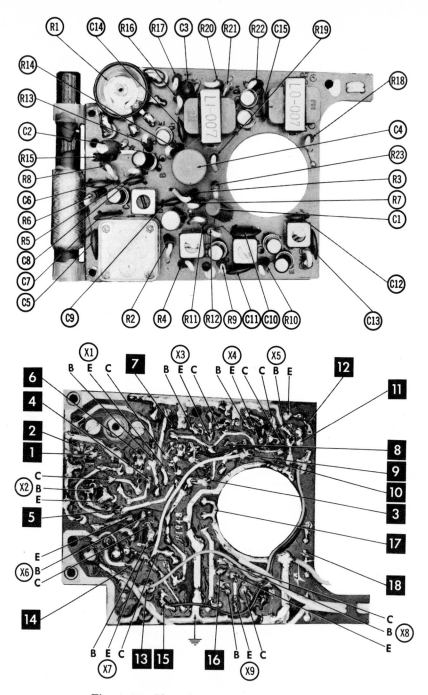

Fig. 4·20 Photodrawings of a radio assembly, both sides.
(A Howard W. Sams Circuitrace Photo.)

are to be performed. Chassis come in various sizes and shapes. Their main purpose is to hold, or support, component parts, but they also often furnish shielding and rigidity and even act as parts of electrical circuits.

4·14 Photodrawing

A fairly new development in the graphics area is photodrawing. Figure 4·20 shows a photograph of an assembled radio with the accompanying letter-number designation of parts. These photographs perform a function similar to that of Fig. 4·18. In some ways a photograph is clearer than a drawing. Certain things, like disk capacitors and coils, show up better in a photograph than in a drawing, especially if the assembly is crowded. Electrical installations that have been extensively modified can be shown better by photographs than by well-worn drawings that have had many changes made on them. Numbers for parts identification, alignment symbols, and connection points are drawn directly on the photograph. A complete list of parts usually accompanies such a photodrawing.

SUMMARY There are many types of production drawings for the construction and assembly of electrical/electronics equipment. Some examples are connection diagrams, panel diagrams, chassis drawings, cabling diagrams, and assembly drawings. Connection diagrams may be classified as point-to-point, airline (baseline), or highway. Careful identification of conductors must usually be made; this may include such items as color, component and terminal destination, size, shielding, and function.

The objects being connected are sometimes shown in symbol form and sometimes in elemental (rectangular or circular) form. Interconnection and cabling diagrams are variations of the connection diagram.

Chassis drawings can be used as the basis for wiring diagrams and parts-assembly drawings. Special problems in dimensioning may arise, requiring rather unique treatment. These drawings are often supplemented by complete parts lists and sets of manufacturing instructions giving each step and tool required. The method of manufacture often dictates how such a drawing is to be dimensioned.

4·1   For what purposes are connection diagrams used?

4·2   What is the difference between an airline, or base-line, diagram and a highway diagram?

4·3   Would it be possible to draw a wiring arrangement as either a baseline or a highway diagram?

4·4   When is it desirable to lay out a wiring diagram in a pictorial (such as isometric or perspective) view?

4·5   If a connection diagram is to show elements (such as tubes, resistors, etc.) of a circuit, to what source would you refer in order to portray those elements?

4·6   Assuming a circuit could be laid out with either a point-to-point or highway type of connection diagram, which type of diagram would probably require more lettering? Why?

4·7   Are connection diagrams sometimes accompanied by other types of drawings or electrical diagrams? If so, what might the other drawing be?

4·8   What letters would you use to indicate the following colors—black, red, white, yellow, grey, brown, green, red-blue?

4·9   In what situations would you use different line widths when drawing a connection diagram?

4·10   What is the sequence or order recommended in ANSI Y14.15 for identification of feed lines in a connection diagram?

4·11   Why is the rear side of a panel usually shown in a panel wiring diagram?

4·12   How is the viewing direction of a chassis layout selected?

4·13   Is it true that airlines (baselines) are usually drawn horizontally? Is it true that highways are usually drawn horizontally?

4·14   It may be said that a cabling diagram is a variation of a certain type of connection diagram. What particular type would this be?

4·15   What are the steps that are used in making a local-cabling diagram?

4·16   What are the general headings of a harness-wire routing list?

4·17   What elements make up a Manufacturing Layout and Time Rate sheet?

4·18 Briefly describe two types of assembly drawings used in assembling various types of electrical or electronics equipment.

4·19 Briefly describe, by sketching, two or more different systems used for the dimensioning of holes on a chassis or board?

4·20 What are some of the objectives that must be borne in mind when one is making an assembly drawing of component parts on a chassis or board?

PROBLEMS 4·1 Redraw the installation shown in Fig. 4·21 as a highway type of connection drawing, with at least three highways. For each lead use standard identification as follows: (1) component destination, (2) terminal,

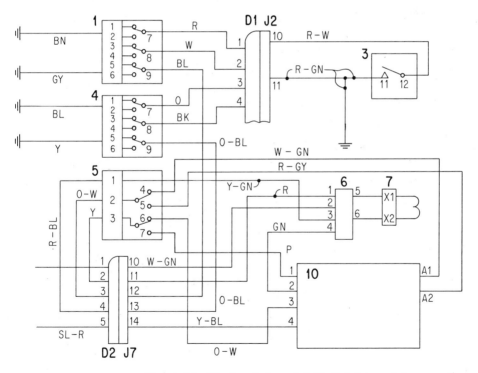

Fig. 4·21 (Probs. 4·1 and 4·2) Point-to-point connection diagram.

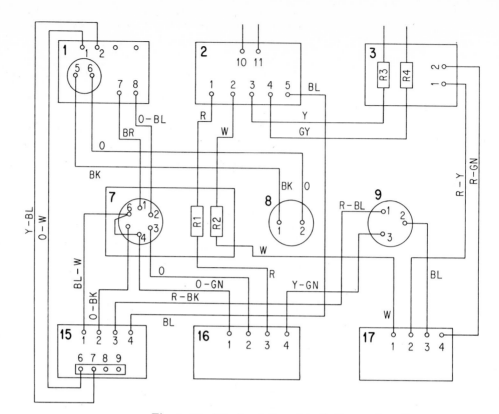

Fig. 4·22   (Probs. 4·3 and 4·4) Connection diagram problem.

and (3) color of wire; e.g., the lead from terminal 7 of component 1 would read D1/1/R and would be placed near component 1. Drawing sheet size: 11 × 17 or 12 × 18. Do not identify leads to grounds, except by color.

4·2   Redraw the installation shown in Fig. 4·21 as an airline-type connection diagram, using several air-lines. Use the standard lead identification suggested in Prob. 4·1, above. Use 11 × 17 or 12 × 18 paper.

4·3   Redraw the connection diagram of Fig. 4·22 as a highway type of connection diagram. Use at least three highways. For each lead, use the standard identification as follows: (1) component, (2) terminal, and (3) wire color. For example, near component 1

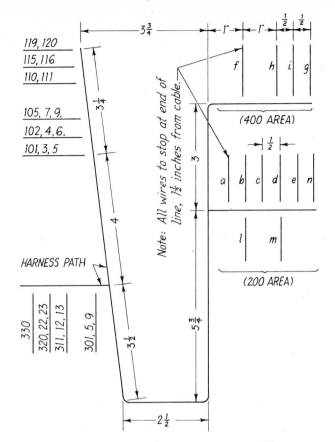

Fig. 4·23 (Prob. 4·5) Cabling harness problem.

the identification of the lead attached to terminal 1 would have the identification 15/6-0-W, meaning that it is attached at the other end to terminal 6 of component 15. Use 11 × 17 or 12 × 18 paper.

4·4 Redraw the connection diagram of Fig. 4·22 as an airline type of connection diagram. For each lead, use the standard designation shown in Prob. 4·3, above. Use 11 × 17 or 12 × 18 paper.

4·5 The rudimentary diagram of a local-cabling harness appears in Fig. 4·23. Complete the diagram using the following information:

start	finish	area
101, 3, 5	f	400 (call 401, 3, 5)
102, 4, 6	e	200
105, 7, 9	d	200
110, 111	c	200
115, 116	b	200
119, 120	a	200
301, 5, 9	n	200
311, 12	i	400
313	m	200
321, 22	h	400
323	l	200
330	g	400

Since starting-line 101 terminates at f in the 400 area, call that 401; where 103 terminates, call that 403, etc. Then add a wire-routing diagram to your drawing. At different places the harness should have different thicknesses which you should estimate.

4·6 Figure 4·24 shows the front, top, and bottom views of a communications satellite package. Make a scale drawing ($\frac{1}{4}'' = 1''$ for 8½ × 11 paper, or $\frac{1}{2}'' = 1''$ for 11 × 17 paper) of these three views, and place view D correctly with regard to these views. Add the cabling to the front view. It is connected through hole B. Add notes for soldering at locations 1, 2, and 3 and for splicing and soldering at location A. Show an appropriate title and your scale.

4·7 The interconnection diagram of Fig. 4·4 is point-to-point. Redraw it as a highway diagram. For the purposes of this problem, cables and other conductors may be merged into the same highway(s) if desired. Use 11 × 17 or 12 × 18 paper.

4·8 Redraw the diagram shown in Fig. 4·4 as an airline

diagram. Use an appropriate method for identifying leads and components. Use 11 × 17 paper.

4·9   Draw the 2″ × 2″ circuit board shown in Fig. 4·25*b* to four times actual size. Then complete the location of the components for the schematic diagram shown in Fig. 4·25*a*. Show the necessary wiring (as hidden lines on the other side of the board) to agree with the schematic diagram. Add notes as required by your instructor. An alternate solution would be to

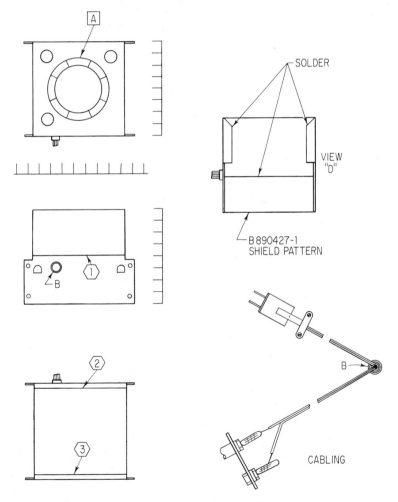

Fig. 4·24   (Prob. 4·6) Satellite oscillator package.

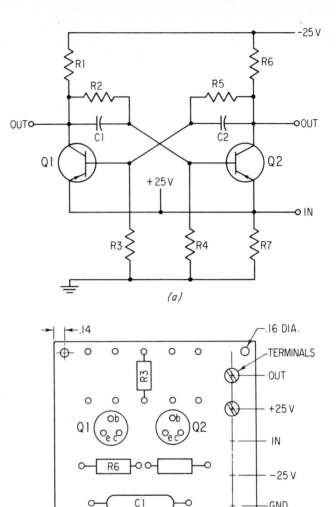

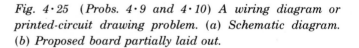

Fig. 4·25 (Probs. 4·9 and 4·10) A wiring diagram or printed-circuit drawing problem. (a) Schematic diagram. (b) Proposed board partially laid out.

draw a mirror image of the board, putting the components in as hidden parts and the wiring in as solid. If it is desired to make a parts list, the values of the components can be found in Fig. 6·6. Use

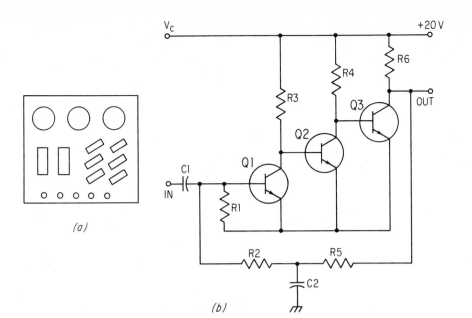

Fig. 4·26 (Prob. 4·11) DC amplifier connection diagram problem.

11 × 17 or 12 × 18 paper. See below for typical component sizes.

4·10 Do the work indicated in Prob. 4·9, but make a printed-circuit-board drawing. (See Chap. 7.) Use 11 × 17 or 12 × 18 paper.

4·11 The circuit shown in schematic form in Fig. 4·26b is to be wired up for a prototype, or experimental, assembly. Using a pattern similar to that shown in Fig. 4·26a, design a component assembly that will fit on as small a card (or board) as possible, leaving about 0.30 clearance from the parts to the edge of the board. Use the ½-watt resistor, small .01 micro-farad capacitor, and .370/.360-diameter transistor shown in the data given on page 104. The wiring should be on the opposite side of the components. If drawn four times actual size, the drawing may fit on 8½ × 11 paper, but it would be safer to use 11 × 17 or 12 × 18 paper.

Typical component sizes

TRANSISTOR DIAMETERS		
a. .370 .360	b. .650 .550	c. .345 .322

RESISTOR SIZES			
	length	*diameter*	*max. resistance*
¼ watt	.375	.093	22 MEG
½ watt	.375	.138	22 MEG
1 watt	.562	.225	22 MEG
1 watt	.715	.237	22 MEG
2 watt	.688	.318	22 MEG

CAPACITOR (TUBULAR PAPER) SIZES		
	200 VDCW	400 VDCW
0.01 microfarad		
small	¼ × ⅝	¼ × ⅝
large	⅜ × 1	⅜ × 1⅛
0.1 microfarad		
small	⁷⁄₁₆ × ⅞	¹⁵⁄₃₂ × 1⅛
large	⁹⁄₁₆ × 1⅝	⅝ × 1⅝

4·12   Figure 4·27 shows a pictorial view and several other partial views of a chassis. The hole sizes are as given in the table at the top of page 106.

Make a complete set of drawings for the construction of this chassis, which has one panel labeled A, two panels on each side marked C (each having two holes), and one panel, labeled B, in front. There is no panel in the rear. Panel B is swung out in the pictorial view to show how C is folded. B is actually vertical. The chassis may be drawn in two ways. Several orthographic views, with dimensions, of the

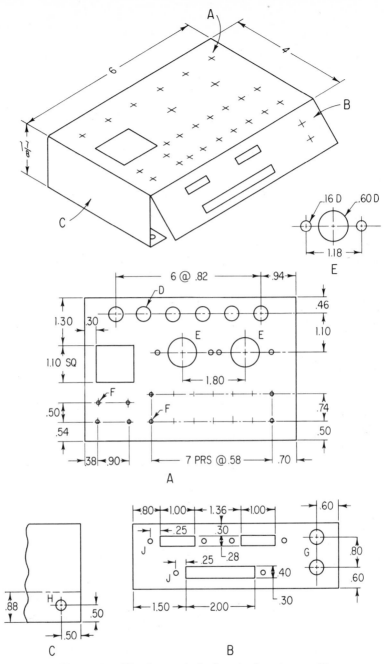

Fig. 4·27 (Prob. 4·12) A chassis drawing problem.

chassis in its completed folded form would be satisfactory. Showing the single piece of metal laid out
flat, with fold lines and a small pictorial view of the

hole	letter or number drill	diameter
D	N	0.302
E	See detail	See detail
F	# 38	.1015
G	K	.281
H	# 27	.144
J	# 20	.161

folded chassis, would also be satisfactory.

All burrs should be removed. The 20-gage steel should be degreased per Spec. 5160 and coated with

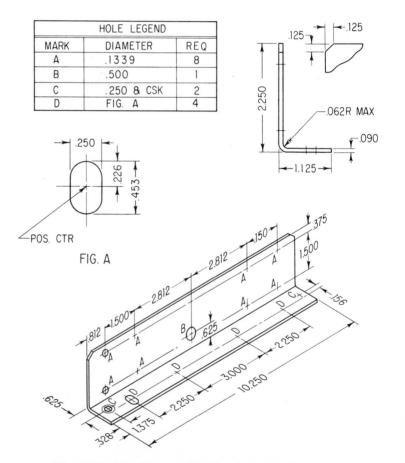

HOLE LEGEND		
MARK	DIAMETER	REQ
A	.1339	8
B	.500	1
C	.250 & CSK	2
D	FIG. A	4

POS. CTR

FIG. A

Fig. 4·28 (Prob. 4·13) Bracket for TD-2 chassis.

clear varnish per Spec. 5160. Tolerance is ±.02 in. Sheet size: 11 × 17, 12 × 18, or larger.

4·13 Figure 4·28 shows a pictorial view and partial views of a bracket, plus a schedule for hole sizes. This bracket is part of the assembly shown in Fig. 4·19. Make a complete drawing for the construction of this part. The bracket is to be made of steel sheet, cold rolled and commercial quality (CRCQ). The holes and ovals are to be punched out and "C" holes countersunk for a .190–32 Flat Head machine screw. The part is to be degreased per 51606 specification and zinc-plated for 289A Finish. Tolerance is ±.016. Use 11 × 17 or 12 × 18 paper.

4·14 Make a three-view drawing of the chassis shown in Fig. 4·29. Members *A* and *C* are identical. The chassis is to be made of 20-gage (.032) CR steel sheet and degreased. Finish is to be gray enamel, baked 525 A, .001 in. thick. Dimension it according to one of the methods prescribed in Prob. 4·12. One of the views may be a flat foldout, or development. This would permit dimensioning for N/C drilling of the chassis. Use 11 × 17 paper.

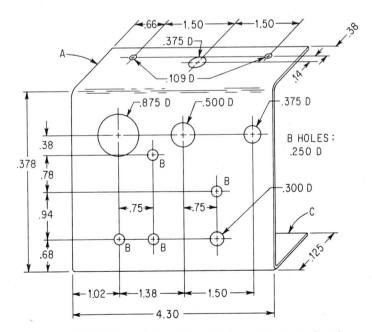

Fig. 4·29 (Prob. 4·14)   Pictorial view of TD-100 chassis.

Chapter 5
Block (flow) diagrams

The block diagram is a useful graphical device in several different areas of electronics and electrical engineering. Because of the simplicity of the block diagram, it is easy to understand the relationships of various parts of a circuit or system if it is presented in this form.

5·1 Examples of block diagrams for electrical circuits and systems

Such a diagram may be used to show the operation of a large electronics system. In such a case, a block would represent a complete and removable chassis, such as a preamplifier, a multivibrator, or even a television camera.

However, in a different situation, a block diagram may be used to facilitate the understanding of a radio receiver or a multistage amplifier, for example. In this case, each block would represent a *stage*. This is the case of the diagram shown in Fig. 5·1. At this time it might be a good idea to define the word *stage*. A stage is considered to be that part of a circuit which includes the main device (e.g., transistor, diode, or tube) and the associated devices that go with it, such as biasing resistors, load resistor, voltage dividers, and capacitors. In other words, a circuit may have several stages, hooked together somehow, so that the signal goes first through one stage, then through the next, and so on. Some of these stages are discussed at a little more length in Chap. 6.

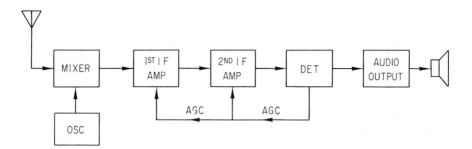

Fig. 5·1 Block diagram of a typical transistor radio receiver circuit.

Figure 5·1 shows how easy it is to understand a circuit's operation by means of a block diagram. It is clearly shown that the signal comes through the antenna (usually portrayed by a symbol rather than a block) and then progresses through the mixer circuit, through the intermediate-frequency (I-F) stages, and finally to the output stage and speaker. The oscillator, which is an auxiliary circuit, is appended to the main circuit; and because it is a frequency generator, its output is fed into the signal train as shown by the arrow. A feedback circuit, labeled AGC for automatic gain control, is correctly drawn below the main circuit. If it were desirable to emphasize the AGC circuit, however, it would be appropriate to place it above the main circuit. The purpose of automatic gain control is to prevent fluctuation in speaker volume when the radio signal at the antenna is fading in and out.

5·2 Preparation of block diagrams

A block diagram may use standard symbols for certain elements, but it is predominantly one of blocks (usually, but not always, squares or rectangles). The layout can be facilitated by (1) using freehand sketches in the initial stages, (2) using cutout cardboard blocks of appropriate size and experimental arrangement until the best pattern is achieved, or (3) using cross-ruled paper for an undergrid, thus facilitating the construction of blocks of equal size and uniform spacing between blocks. The size of the rectangles is usually determined by the lettering that goes in them; and since they are usually drawn about the same size, the block with the most lettering will often set the size of the blocks in the entire diagram.

From Fig. 5·1 and other diagrams in this chapter, certain facts about block diagrams can be deduced, and the following rules for their construction can be listed:

1 The signal path should be made to go from left to right, if possible. In large, complex drawings, the input should preferably be at the upper left and the output at the lower right, if possible.

2 Blocks are usually drawn in one of three shapes: rectangular, square, or triangular. (The triangle represents different items in different types of diagrams. There are also other shapes for certain specialized diagrams, as will be shown later in the chapter.)

3 Once the size and shape of a block is determined, the same size and shape should be used throughout the drawing. The size of a rectangle, for instance, bears no relation to the importance of the component(s) it represents.

4 A single line, preferably heavy, should be used to show the signal train from block to block. In complex circuits or systems, however, more than one line may have to be drawn leading into or away from a block.

5 Arrows should be used to show the direction of signal flow.

6 Some components, usually terminal ones such as antennas

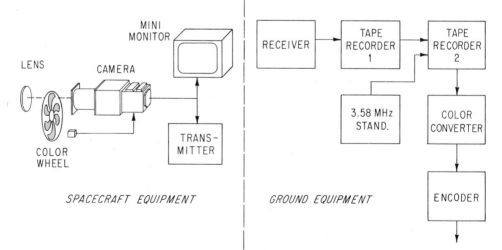

Fig. 5·2 *Block diagram of the Apollo color television system.* (*Westinghouse Electric Corp.*)

and speakers, are shown by means of standard symbols rather than by blocks.

7 Titles, or brief descriptions, of the components or stages represented should be placed within the blocks.

Aside from the above-listed rules, no standardized procedure exists for the preparation of block diagrams. In Fig. 5·1, for instance, either square or rectangular blocks could have been used. The arrows, which are shown touching the blocks, could have been placed midway between the rectangles if desired.

In Fig. 5·2, both the spacecraft and ground systems are shown. Part of the camera's apparatus is shown pictorially. (It could have been drawn as blocks.) It generates a field-sequential color signal using a rotating color wheel and a single-image tube. The tape recorders compensate for Doppler shift and present real-time information to the scan (color)

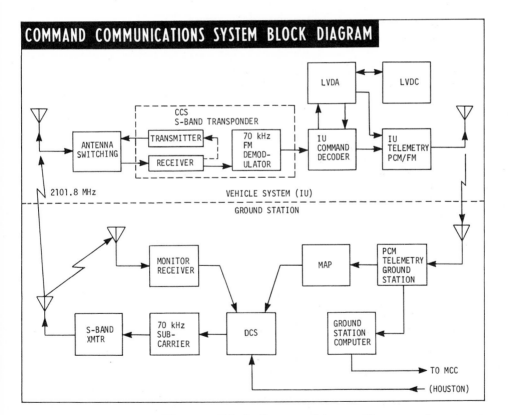

Fig. 5·3 Block diagram of Saturn V command communications system. (NASA.) LVDA, launch vehicle data adopter; CCS, command communication system; MAP, message-acceptance pulse; and DCS, digital computer system.

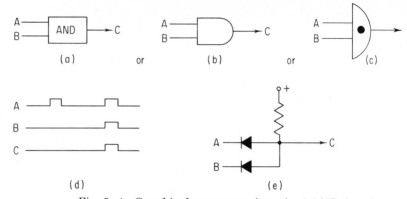

Fig. 5·4   Graphical representation of an AND function or circuit at (a) (b) and (c). (d) Impulses occurring at different (left) and the same (right) instances. (e) An AND circuit.

converter. The second tape recorder is driven by a standard frequency to correct any tape-speed errors. The converter is a storage and readout device.

Figure 5·3 shows the command communication system, which provides for digital data transmission from ground stations to the LVDC (launch vehicle data converter). The instrument unit (IU) is a structure located on top of the SIV-B stage of Saturn V. It contains the guidance, navigation, and control equipment which guide the vehicle into its mission trajectory.

5·3 Logic diagrams

The next few figures will show some of the main functions of switching circuits. These functions can be shown by appropriate blocks in block or flow diagrams. Most of these functions and their graphical representation can be found in ANSI Y32.14, "Graphical Symbols for Logic Diagrams."

The first function is the AND function. It can be depicted graphically several ways, as shown in Fig. 5·4. Figure 5·4a to c shows the approved blocks for use in block diagrams. The rectangular block must carry on A or AND, while the other blocks are left unlettered inside. (More than two input leads may enter each block.) If a pulse of, say, 5 volts goes into input A and at the same instant a similar pulse enters input B, there will be a definite output pulse at C. If, however, a pulse enters A but no pulse goes into B, there will be no output pulse. Another way of showing the characteristics of an AND circuit is the *truth table*. The following truth table

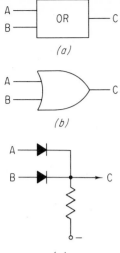

(a)

(b)

(c)

Fig. 5·5 Graphical representation of an OR function or circuit.

A	B	C
0	0	0
0	1	0
1	0	0
1	1	1

The AND function

It shows that an output will result only when all inputs receive impulses at the same instant. (The output assumes the 1-state if and only if all the inputs assume the 1-state.)

The next function is the OR function. In this circuit, the output has a significant pulse if there is a significant pulse at either A or B or both. Or, to use the language of the engineer: The output assumes the 1-state if one or more of the inputs assume the 1-state. The positive truth table would read:

A	B	C
0	0	0
0	1	1
1	0	1
1	1	1

The OR function

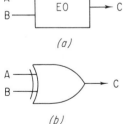

(a)

(b)

Fig. 5·6 Block diagram symbols for the EXCLUSIVELY OR function as approved by ANSI Y32.14.

The next function is the EXCLUSIVELY OR function. As the truth table shows, this function differs from the OR function in that impulses at both A and B will not produce an output pulse.

A	B	C
0	0	0
0	1	1
1	0	1
1	1	0

The EXCLUSIVELY OR function

The next function is the NOT function. It says, in effect,

that C is not equal to A. It is also referred to as an *inverter*.

A	C
0	1
1	0

The NOT function

The next circuit is the NOR circuit. It says, in effect, that C is *not A or B*. NOR, therefore, is a contraction of the words *not* and *or*.

A	B	C
0	0	1
0	1	0
1	0	0
1	1	0

The NOR function

The next function is the NAND function. This means that C is *not A and B*. C is in the 0-state if all the inputs are in the 1-state, and C is in the 1-state if one or more inputs are in the 0-state.

A	B	C
0	0	1
0	1	1
1	0	1
1	1	0

The NAND function

The American National Standard (Y32.14) does not specify any specific shapes for the NOT, NOR, and NAND functions, other than permitting the use of the rectangular or square box. Therefore, unless one wishes to design special shapes for these functions, he will show them with a box and the appropriate letters, NOT, NOR, or NAND inside. It should be noted that *inputs can be drawn on any side of the symbol except the output side.*

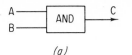

(a)

A ———o| (AND) |o——— C
B ———o|

(b)

Fig. 5·7 An AND *circuit using positive logic at (a) and negative logic at (b). The result of using negative logic is that the* AND *circuit performs the function of an* OR *circuit.*

5·4 Negative and mixed logic

Figure 5·7b shows an AND circuit with small circles at the input and output junctions. These circles indicate that negative logic is being used. The negative truth table would look like this:

A	B	C
1	1	1
1	0	1
0	1	1
0	0	0

Negative truth table for AND function*

A glance at the truth table for the OR function, shown earlier, will reveal that these two truth tables are the same. Therefore, if negative logic is used, the AND circuit performs the same function as an OR circuit performs in positive logic. Similarly, the other circuits previously described perform different functions as follows:

circuit or function	function performed under negative logic
OR	AND
NAND	NOR
NOR	NAND

However, the NOT circuit performs as usual, whether negative or positive logic is used.

The use of negative or mixed logic has some advantages in computer design. Some computers utilize negative logic in order to achieve more economical circuitry. One specific advantage is that by using negative and positive logic, just two types of circuits—for example, NAND and NOR—can be used to perform most of the needed work of the computer. Figure 5·8 shows a circuit drawn with two types of logic diagrams. Figure 5·8a utilizes uniform shapes for functions, and Fig.

*Compared with the AND table, shown earlier, each pulse or state (both input and output) is just the opposite.

$5 \cdot 8b$ uses distinctive shapes. The truth table lists only a few of the 16 possible input combinations. Note the small circles which indicate the use of negative logic at those locations.

Table 5·1   Decision table for an automated rapid-transit system

rules	where event occurred*	is train approaching CP at ≤ minimum tolerance?	will first-come– first-served give correct sequence?	will train behind be delayed?	train perfor- mance index†	can trains ahead be slowed?	actions‡
1	0	Yes	Yes	Yes			2
2	0	Yes	Yes	No			0
3	0	Yes	No				3
4	0	No		Yes			2
5	0	No		No			0
6	1			Yes			2
7	1			No			0
8	2				0		0
9	2				1		0
10	2			Yes	2		2
11	2			No	2		0
12	2			Yes	3		2
13	2			No	3		0
14	2				4	Yes	1
15	2				4	No	4

*State 0, between a station and a "merge"; state 1, between a station and a CP that is not a merge; state 2, at least one station before a CP.

†State 0, < 10 sec late; state 1, 10 to 30 sec late; state 2, 30 to 60 sec late; state 3, 60 to 120 sec late; state 4, < 120 sec late.

‡Action 0, continue with existing schedule; action 1, revise schedules ahead of delayed train to reduce extended gap; action 2, revise schedules behind delayed train to extend reduced gap; action 3, recommend revised sequence at interlocking; action 4, recommend station run-through.

5·5 The decision table

Somewhat similar to the logic table is the decision table, which is essentially a tabulation of logical relationships consisting of conditions, actions, and rules. Conditions are the variables that influence any decision, while actions are the things to be done once a decision has been made. In the case of BART's highly automated rapid mass-transit system (Fig. 5·9 and Table 5·1), a computer makes the decision and one of five actions (0 to 4 listed at the bottom of the table) is taken. A CP is a critical point (control location) at which it is especially desirable for trains to be on time.

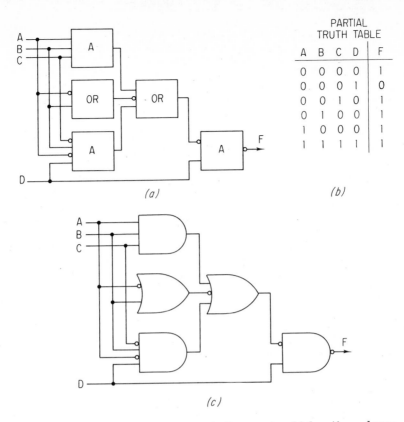

A	B	C	D	F
0	0	0	0	1
0	0	0	1	0
0	0	1	0	1
0	1	0	0	1
1	0	0	0	1
1	1	1	1	1

(a)

(b)

(c)

Fig. 5·8   Example of a logic diagram in which uniform shapes have been used at (a) and distinctive shapes at (c).

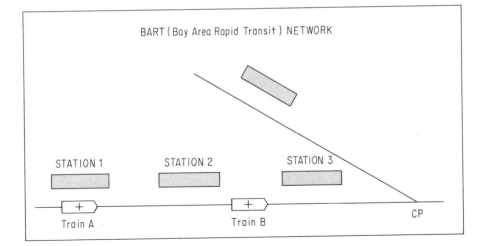

BART (Bay Area Rapid Transit) NETWORK

STATION 1

STATION 2

STATION 3

Train A

Train B

CP

Fig. 5·9   Simplified diagram of the BART network illustrates how the decision-table logic matches system conditions with a rule (in this case rule 14.) Rule 14 specifies action 1, which is to revise schedules ahead of delayed train to reduce the gap.

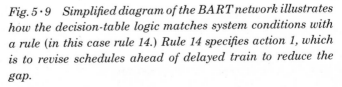

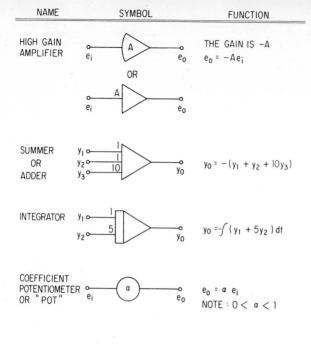

HIGH GAIN
AMPLIFIER

THE GAIN IS $-A$
$e_o = -Ae_i$

OR

SUMMER
OR
ADDER

$y_o = -(y_1 + y_2 + 10y_3)$

INTEGRATOR

$y_o = -\int (y_1 + 5y_2) \, dt$

COEFFICIENT
POTENTIOMETER
OR "POT"

$e_o = a \, e_i$
NOTE : $0 < a < 1$

Fig. 5·10 Basic functions of general purpose analog computers and their symbols.

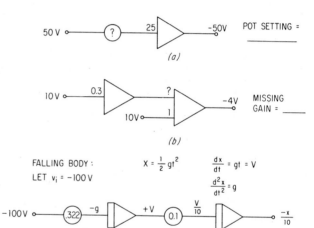

50 V   ?   25 $-50V$ POT SETTING = _____

(a)

10 V   0.3   ? $-4V$ MISSING GAIN = _____

10V   1

(b)

FALLING BODY :
LET $v_i = -100 \, V$

$X = \frac{1}{2}gt^2$ $\frac{dx}{dt} = gt = V$

$\frac{d^2x}{dt^2} = g$

$-100V$ .322 $-g$ $+V$ 0.1 $\frac{V}{10}$ $\frac{-x}{10}$

(c)

EQUATION : $2.1 \, x' - 4.72 \, x''' = x''$

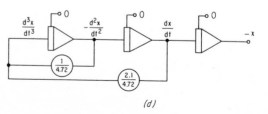

$\frac{d^3x}{dt^3}$ $-\frac{d^2x}{dt^2}$ $\frac{dx}{dt}$ $-x$

$\frac{1}{4.72}$ $\frac{2.1}{4.72}$

(d)

Fig. 5·11 Four problems set up by diagrams for solving on an analog computer.

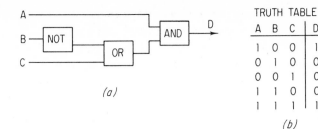

TRUTH TABLE

A	B	C	D
1	0	0	1
0	1	0	0
0	0	1	0
1	1	0	0
1	1	1	1

(a)

(b)

5·6 Analog-computer programming diagrams

An important step in the writing of a program to be solved on a computer is the making of a flow diagram. Certain standardized blocks are used in making these diagrams. Figure 5·10 shows the basic blocks for analog-computer diagramming. Not shown are such functions as the servo multiplier, division circuit, diode function generator, and servo function generator.

These are the things that can be done with a medium-size, modern, general-purpose analog computer:

1 Add
2 Subtract
3 Multiply:
   a. Easily by a constant
   b. Not so easily by a variable
4 Integrate
5 Generate functions
6 Display or record the results:
   a. By a meter
   b. By an oscilloscope
   c. By a recorder or plotter

The upper limit of the *gain* of good high-gain amplifiers is 10^8. A gain of 5, for example, means that the output signal is five times as strong as the input signal. One problem facing the programmer is that of scaling the output so that it will be properly recorded. In other words, if he uses voltages that are too large, the curve representing the answer will be too large to fit on the recording graph or oscilloscope. Or if he uses voltages that are too small, his answer curve will be too flat to give an accurate answer. Available working voltages are generally between -100 volts and $+100$ volts on large- and medium-size computers, and -10 volts and $+10$ volts on small analog computers.

Figure 5·11 shows the flow diagrams which would be used for solving four problems on the analog computer. The first two are quite simple. Problem a requires the correct potentiometer setting to achieve an output of −50 volts. The answer is 0.04. Problem b requires the correct gain to place at the upper input to the second adder to achieve the indicated output. The answer is 2. (The output of the first adder is −3 volts.) Problem c involves two integrators and pots. If the input of −100 volts, later multiplied by 0.322, represents d^2x/dt^2, then the output of the first integrator is dx/dt, which is the velocity. For scaling purposes, a second potentiometer is set to 0.1, and the integral of $v/10$ turns out to be $−x/10$. Now, if we get an output of 10 volts $(x/10)$, $x = 100$ and our answer is 100 ft. Problem d involves three integrators in order to solve for x. This diagram shows how the coefficients in the equation can be set up with potentiometers. It also shows initial conditions of zero set into each integrator. Usually, but not always, the initial conditions are zero.

The author realizes that many students who use this text may not have had much calculus instruction. He also realizes that there is more to analog computer programming than what has been mentioned here. Therefore, the reader may not understand all the problems discussed here. But all readers should appreciate the importance of the flow diagrams in the solving of such problems.

5·7 Digital computer programming diagrams

Engineers, scientists, and certain technicians in the electrical and electronics fields may find it advisable to use the analog computer for certain problems and the digital computer for other problems. Figure 5·12 shows the basic blocks for flow diagrams in digital programming. We believe that two examples will show fairly clearly how the boxes are used.

The problem shown in Fig. 5·13a is that of obtaining the shortest distance between two points in space, given any number of x, y, and z coordinates for the two points. The problem is programmed so that two sets of coordinates will be read, the distance computed and then printed (on tape, typewriter, or a punched card), and the same procedure repeated until there are no more coordinates left. This is shown on the decision block, labeled $n:0$. If n is equal to zero (Yes), the program is finished; but if n is not equal to zero, there are more coordinates yet to be read and the computer will compute D for the next set of coordinates.

Problem b consists of solving p for N values or combinations of a, b, and c. It has two decision boxes, one for the

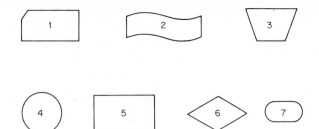

Fig. 5·12   Basic symbols for flow diagrams used in digital computer programming. (1) Punched card. (2) Tape. (3) Input, output. (4) Magnetic tape or termination. (5) Processing, computation. (6) Decision (branching). (7) Start or stop.

same purpose as the previous problem. But the first box, labeled $C:0$, is to take care of the situation when a C value should happen to be zero. A glance at the argument indicates that p will be infinity if C equals zero. The computer will try to go to infinity, and precious minutes or hours will be wasted because it cannot reach infinity. To avoid this possibility, the computer can be instructed to do something else whenever a C value of zero shows up. In this case, the instruction is

Fig. 5·13   Flow diagrams for two prob-lems programmed for solution on a digital computer.

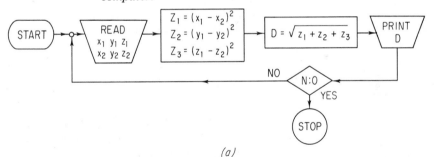

(a)

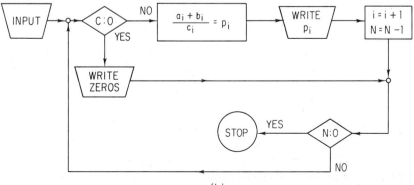

(b)

to punch or print a series of zeros, then start over with a fresh set of values.

Flow diagrams may be drawn from left to right and possibly from right to left if there is a second row of blocks below the first. Or they may be drawn from top to bottom. The function boxes are sometimes drawn the same size throughout a diagram, but because of the varying amounts of material that may be placed in these boxes, they are often drawn in various sizes and proportions.

SUMMARY Flow diagrams are used for different purposes in different situations. They may be used for depicting an electrical circuit or system, for the preliminary design work in computers and other electrical installations, and for programming problems to be solved by computers. A left-to-right direction is sought in planning most block diagrams because this is the normal way in which people read. However, this sequence cannot always be followed. In most block diagrams there is a certain amount of lettering within the blocks. This lettering may determine what the sizes of the blocks will be. Liberal use of arrowheads is made in block-diagram construction. Although the rectangle is widely used in flow and block diagrams, other shapes—and sometimes electrical symbols—are used as dictated by standard practice.

QUESTIONS

5·1 When would you use a block diagram to show the electrical properties of a packaged circuit?

5·2 In making a block diagram of an electrical system, what shapes would you use for the blocks? Why?

5·3 What direction of flow would you attempt to show in planning a block diagram of an electronic circuit?

5·4 Sketch six different shapes that may be used at one time or another in flow diagrams. Label each shape.

5·5 What are five different functions that might be shown with the rectangle in block or flow diagrams?

5·6 What are three devices that may be shown by means of standard symbols, rather than by boxes, according to customary practice?

5·7 Define the word *stage*.

5·8   What may a triangular "block" represent?

5·9   Where are auxiliary circuits, such as feedback, usually placed with regard to the main signal train?

5·10   What standard covers the preparation of block or flow diagrams?

5·11   What is the difference between an EXCLUSIVELY OR circuit and an OR circuit? (Show by means of a truth table.)

5·12   Add the missing lines in the truth table of Fig. 5·8b.

5·13   Add the missing lines of the partial truth table of Fig. 5·9b for A and B only, in the 1-state; A and C only; and A and D only, in the 1-state.

5·14   What are the advantages of using distinctive shapes, rather than uniform shapes, in a logic diagram? A disadvantage?

5·15   Is a positive voltage signal which is input into an adder converted to a negative voltage output?

5·16   Is a negative voltage which is input into an integrator converted to a positive voltage output?

5·17   What are the upper and lower limits of a coefficient potentiometer in most analog computers?

5·18   What voltage limits are available in most analog computers?

5·19   Why is a decision box sometimes called a *branching box?*

5·20   Are the function boxes in a digital-computer flow diagram usually drawn to the same size? Why?

PROBLEMS   5·1   Make a block diagram for the TR605 radio receiver circuit as follows: Use a left-right sequence with arrowheads on or above the lines connecting the blocks. Make neat uppercase lettering in the blocks, using appropriate abbreviations: (1) Converter, (2) 1st I-F Amp, (3) 2nd I-F Amp, (4) Driver, (5) Output stage. Use an antenna symbol for input and a speaker symbol for output. Use 8½ × 11 paper.

5·2   Construct a block diagram for the following (T-50A) radio circuit (follow the instructions contained in Prob. 5·1):

a. Antenna
b. Converter
c. 1st I-F Amp
d. 2nd I-F Amp
e. 1st A-F Amp
f. 2nd A-F Amp
g. Output stage
h. Speaker or earphone

Sheet size: 11 × 17 or 12 × 18 (or on 8½ × 11 paper if drawn as two lines).

5·3 Make a block diagram for the following (TR 9–10 BC-SW) radio receiver (follow the instructions given in Prob. 5·1):

a. External antenna
b. Mixer
c. 1st I-F Amp
d. 2nd I-F Amp
e. A-F Amp
f. Driver
g. Output stage
h. Earphone or speaker jack
i. Oscillator to feed into the mixer; *and* AGC feedback around the I-F stages

Use 11 × 17 or 12 × 18 paper unless drawn as two lines, in which case it might fit on 8½ × 11 paper.

5·4 Make a block diagram for the following FM radio receiver (see instructions given in Prob. 5·1):

a. Antenna
b. FM RF Amp
c. FM Converter
d. 1st FM I-F Amp
e. 2nd FM I-F Amp
f. 3rd FM I-F Amp
g. A-F Amp
h. Output stage
i. Jack

Use 11 × 17 or 12 × 18 paper.

5·5 Arrange the following parts of a digital computer into five blocks with the control unit above the memory unit, both being in the center part of the diagram: (1) Control unit, (2) Memory (storage), (3) Arithmetic, (4) Input-output, (5) Control console. Then, in medium lines (or red lines), draw the following lines for control functions: Control console to control unit, and vice versa. Control unit to Input-output. Control unit to Memory, and Control unit to Arithmetic. Then, with heavy lines (or black lines), draw the following paths for data flow: Memory to control. Memory to Input-output and vice versa. Memory unit to Arithmetic and vice versa. Use 8½ × 11 paper.

5·6   Draw in left-to-right direction the following steps in computer programming for numerical controlled machinery:

a. Part program
b. Card punch
c. Program deck
d. Computer
e. Magnetic tape
f. Conversion equipment
g. Punched tape
h. Machine tool (control)

Also, going into the computer draw:

i. Computer program
j. Postprocess program

Rectangular blocks may be used, or pictorial symbols, such as shown in Fig. 5·14. Use 11 × 17 or 12 × 18 paper.

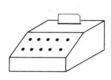
PART PROGRAM

MAGNETIC TAPE

PUNCHED TAPE

CARD PUNCH

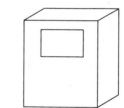
CONVERSION EQUIP'T

PROGRAM DECK
COMP. PROGRAM
POSTP. PROGRAM

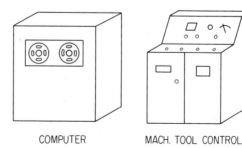
COMPUTER   MACH. TOOL CONTROL

Fig. 5·14   (Prob. 5·6) Pictorial symbols for use in Prob. 5·6.

5·7 Construct a block diagram for a basic regulating system that has the following steps or devices:

a. Power source b. Regulated quantity
c. Signal-sensing device d. Error-sensing device
e. Reference f. Amplifier with
g. Regulator power source feedback

Such a regulating system could be used for the speed control of a motor. In such a case, the motor would be the regulated quantity and a tachometer would be the signal-sensing device. Use 8½ × 11 paper.

5·8 Make a block diagram of the post amplifier for the Apollo TV camera shown in Fig. 5·15. Place the fol-

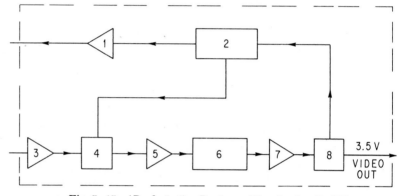

Fig. 5·15 (Prob. 5·8) Post amplifier for Apollo TV camera. (Westinghouse Electric Corp.)

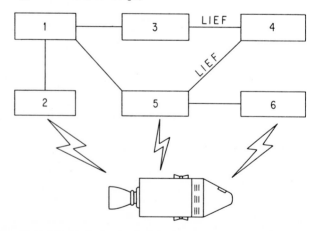

Fig. 5·16 (Prob. 5·9) Telemetry system for Apollo–Saturn spaceflight. (NASA.)

lowing lettering in or near each block indicated:
(1) Amp (amplifier), (2) Detector and Discriminator,
(3) Amp, (4) AGC, (5) Amp, (6) Aperture control,
(7) Amp, (8) Mixer.

5·9 Complete the block diagram of the telemetry command and communication interfaces for Apollo–
Saturn flight control shown in Fig. 5·16. Add the
following lettering at the places indicated by the
numbers: (1) Goddard, (2) Manned Spaceflight Network (MSFN), (3) Houston, (4) Marshall, (5) Kennedy,
(6) AF eastern test range (AFTR). Add a suitable title
and the letters LIEF (launch information exchange
facility) where indicated. Use 8½ × 11 paper.

Fig. 5·17 (Prob. 5·10)
Indicator block dia-
gram for airborne
radar.

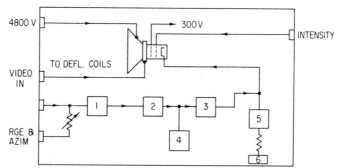

5·10 Complete the flow diagram displayed in Fig. 5·17
by adding the following titles to the boxes indicated
by the numbers: (1) Video mixer, (2) Video amplifier,
(3) Video amplifier, (4) DC restorer, (5) DC restorer,
(6) Ground (symbol).

5·11 Figure 5·18a to f shows simple flow diagrams for
analog computer problems. Supply the missing information as follows:

a. output ____   b. input ____
c. pot setting ____   d. missing gain ____
e. output ____   f. input ____

Fig. 5·18 (Prob. 5·11)
Analog computer flow
diagrams.

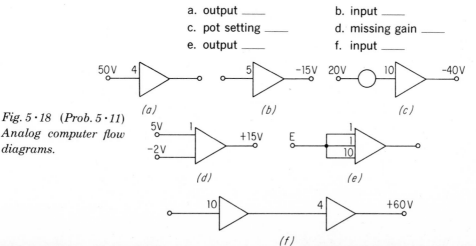

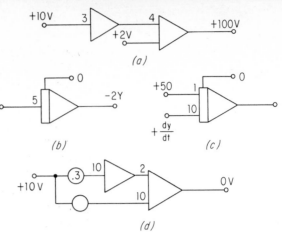

(a)

(b)

(c)

(d)

5·12 (Parts of this problem require a fair understanding of calculus.) For the diagrams shown in Fig. 5·19, supply the missing information as follows:
a. missing gain ____   b. input ____
c. output ____   d. pot setting ____

5·13 Figure 5·20 shows a flow diagram for a digital computer program to solve for x and y using Cramer's rule. If $ax + by = p$ and $cx + dy = q$, analyze the program mathematically. Then analyze the program step by step, with a sentence or two describing each block. Can you add something to the flow diagram to make it stop after all values of a, b, c, and d have been read into the computer? (As written, this is not very clear.)

5·14 Write a flow diagram for a digital computer solution for values of e^x for $x = 1$ through 20. $e^x = 1 + x/1! + x^2/2! + \ldots x^n/ n!$.

5·15 Write a flow diagram for one of the following situations.
a. The area of a trapezoid
b. The area under a curve, using trapezoidal or Simpson's rule
c. The area and volume of a sphere
d. The sum of all consecutive numbers from 1 to 500

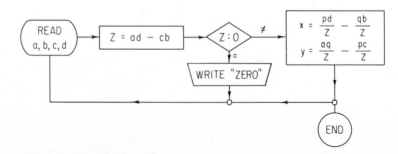

Chapter 6

The elementary, or schematic, diagram

The elementary diagram—also often referred to as the *schematic diagram*—is the hallmark of the design room of the electronics industry. It is the diagram that shows the functions and relations of the component devices of a circuit by means of graphical symbols. It does not show the physical relationship of those components, however.

In the communications field, it is usually referred to as the *schematic diagram;* in the electrical controls area, it is called the *elementary diagram.*

Such a diagram makes it possible for a person schooled in electronics to trace a circuit with comparative ease. For this reason it is used for design and analysis of circuits, for instructional purposes, and for trouble-shooting. The elementary diagram is also used in other electrical areas which are not strictly electronic. One such area is the electric-power field, discussed in Chap. 9.

6·1 Examples of transistors in circuit drawings

In order to give the student a "feel" for the layout of schematic diagrams, we shall present several figures showing popular formats now in use. Then the problem of laying out such a diagram will be discussed. (A student without much background in electronics may not be able to understand all the author's comments about the following circuits, but he should be able to visualize the different drawing patterns that are shown.)

Figure 6·1 shows the three accepted methods of connecting a transistor in a circuit: *common base* (or grounded base), *common collector,* and *common emitter.* In the common-base circuit, Fig. 6·1a, for example, the signal (shown by the ~ in the circle, which indicates an a-c input) is introduced into the emitter-base circuit and extracted from the collector-base circuit. The base is thus common to both the input and output circuits. The direction of the arrows shows the electron flow. The voltage or power gain may be in the order of 1,500, and the phase of the signal is not changed.

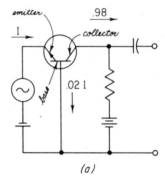

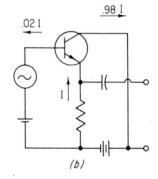

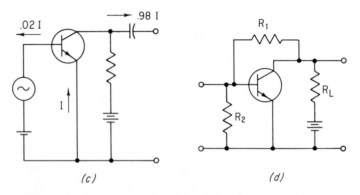

Fig. 6·1 *Basic transistor circuits. (a) Common base. (b) Common collector. (c) Common emitter. (d) Bias network for common-emitter circuit. (From RCA Transistor Manual.)*

In the common-collector arrangement, Fig. 6·1b, the signal is introduced into the base-collector circuit and extracted from the emitter-collector circuit. The power gain is lower than in the other two configurations, and there is no phase reversal. This arrangement is used primarily as an impedance-matching device.

The common-emitter circuit, Fig. 6·1c, can provide power gains of 10,000. The input signal is introduced to the base-emitter circuit, and the output is taken from the collector-emitter circuit. The output signal voltage is 180° out of phase with the input signal. However, this arrangement is the most widely used when more than one stage is required.

Figure 6·1d shows a popular biasing arrangement for a common-emitter circuit, which does away with one of the batteries used previously. (Bias is the difference in potential between, say, the collector and the base.) A voltage-divider network composed of $R1$ and $R2$ provides the required forward bias across the base-emitter junction.

In the examples shown, NPN transistors have been used. PNP transistors could be used instead, in which case the battery polarities should be reversed.

6·2 The basic amplifier

Three major functions of transistors and electron tubes are *amplification, oscillation,* and *switching.* Four different *classes* of amplifier service are as follows:

Class A The collector (or plate in a tube) current flows continuously during the complete electrical cycle.

Class AB The collector or plate current flows for appreciably more than half the cycle but less than the entire cycle.

Class B The collector or plate current flows for approximately one-half of each cycle when an alternating signal is applied.

Class C The collector or plate current flows for considerably less than one-half of each cycle when an alternating signal is applied.

Figure 6·2a shows a basic class A amplifier circuit with a PNP-type transistor. Such amplifiers are used in low-level audio stages, such as preamplifiers and drivers, where "noise" will be at a minimum. Resistors $R1$ and $R2$ determine the base-emitter bias, $R3$ is for emitter stabilization, and the output signal is developed across the collector-load resistor R_L. Capacitor $C1$ bypasses the a-c signal around $R3$. (A *loop* has been shown in Fig. 6·2a between $R3$ and the transistor to emphasize the fact that there is no connection at this location. Such loops are not standard and, as a rule, will not appear in drawings in this book. The author elected to use loops in this figure and in Fig. 6·1b. From now on, except in one or two cases, the standard method of showing crossing, but not connecting, lines will be used.)

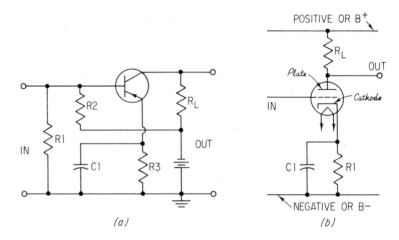

Fig. 6·2 *Basic amplifier circuits.* (a) *Transistor* (*triode*). (b) *Vacuum tube* (*triode*).

A similar amplifier circuit is shown with a triode vacuum tube in Fig. 6·2b. Here the cathode is kept positive by means of the biasing resistor $R1$ and the bypass capacitor $C1$. The capacitor keeps the voltage steady regardless of changes of tube current. R_L is the load resistor.

6·3 Interstage coupling

In the main, there are three methods by which two or more stages (amplifier stages, for example) may be hooked together. These three methods are (1) RC (resistance-capacitance), (2) transformer, and (3) DC (direct coupling). Each has its advantages and disadvantages. Figure 6·3 shows examples of these methods.

Fig. 6·3 Methods of coupling amplifier stages together. (a) Part of an RC coupled amplifier. (b) Transformer coupling. (c) Direct-coupled (DC) shunt regulator circuit.

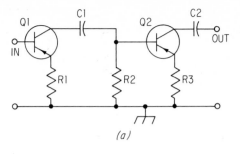

(a)

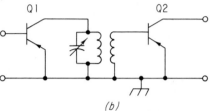

(b)

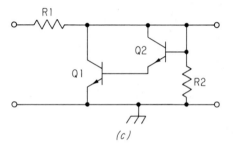

(c)

Figure 6·3a shows part of a resistance-capacitance-coupled amplifier. C1 is called the coupling capacitor. This method is widely used because of its low cost and the large range of frequencies that can be handled.

Figure 6·3b shows a transformer-coupled network in which one side can be "tuned." (Note that in Fig. 6·3b and c the transistor symbols have not been enclosed by an envelope circle. The reader may recall from Chap. 3 that the circle is optional. The author has omitted the circle from these two circuit drawings on purpose to remind the reader that it is not always used. The author prefers the circle, however, and uses it in most examples in the book.) Transformer coupling may also be employed in which both sides or neither side is tuned. Tuning makes possible frequency selection. The use of transformers in coupling has several advantages, but it also is expensive.

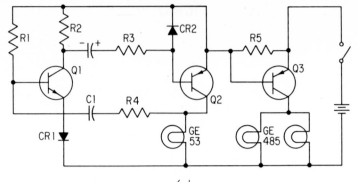

(a)

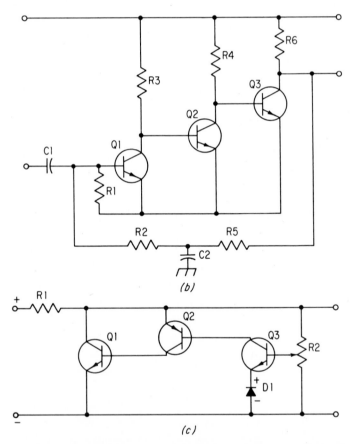

(b)

(c)

Fig. 6·4 *Different patterns of transistors in circuits. (a) Highpower light-flasher circuit. (b) Common-emitter DC amplifier. (c) Shunt regulator circuit. (Figures (a) and (b) from GE Transistor Manual. Figure (c) from Milton S. Kiver, "Transistors," 3d ed., McGraw-Hill Book Company, New York, 1962. Used by permission.)*

Figure 6·3c shows a direct-coupled circuit, which regulates the circuit so that the output voltage is maintained nearly constant. *DC* amplifiers are normally used to amplify small d-c or very-low-frequency a-c signals. Typical applications include output stages of series-type and shunt-type regulating circuits, chopper circuits, and differential and pulse amplifiers. Obviously, the cost is low for this type of coupling, and it is, therefore, widely used.

6·4 Patterns for transistor circuits

The nature of the circuitry, including coupling circuits, often indicates a pattern to the engineer or draftsman who is making the final sketch or layout drawing of an electronics circuit. He can plan his drawing to allow the pattern to develop in as auspicious a manner as possible.

Figure 6·4a shows a flasher circuit for emergency vehicles, barricades, boats, and aircraft. It will be noted how the transistors are aligned on a horizontal line (invisible), as are the flasher lamps and some other components of the circuit. This type of transistor alignment is possible in a good many circuits, radio and television among them.

Figure 6·4b shows another pattern which develops because of the direct coupling of the transistors. The second and third stages are biased by the preceding stages. A feedback loop containing $R2$, $R5$, and $C2$ adds additional stability to the circuit.

Figure 6·4c shows another pattern which develops when, for instance, PNP and NPN transistors are used in a *DC* arrangement. A shunt regulator circuit is used to regulate a power-supply *output voltage*. This circuit has a corrective process which ensures the same output regardless of increases or decreases in voltage from the power supply.

Another arrangement of transistors is shown in Fig. 6·5. In a push-pull circuit, such as that shown in Fig. 6·5a, each transistor amplifies half the signal, and these half-signals are then combined in the output (collector) circuit to restore the original waveform in an amplified state. This circuit requires transformer coupling, while that of Fig. 6·5b does not. In this circuit, showing electron current flow by means of arrows, essentially no d-c current flows through resistor R_L. Therefore the voice coil of a speaker can be connected in place of R_L without excessive speaker-cone distortion.

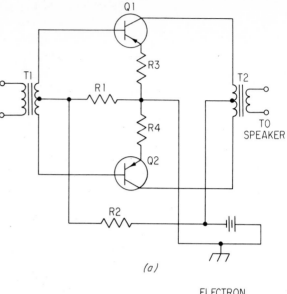

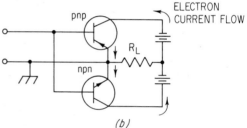

Fig. 6·5 Other patterns. (a) A push-pull amplifier circuit. (b) Basic complementary-symmetry circuit. (From RCA Transistor Manual.)

6·5 Examples of computer circuits

Figure 6·6a shows a saturated flip-flop circuit having one input, called a *trigger,* at T and outputs at A and B. A flip-flop is a memory, or storage, device that may have one of two states, ON or OFF. It always has two complementary level outputs and may have one, two, or three inputs.

A common feature of flip-flop circuit drawings is the crossing, angled, signal-path lines near the center of the circuit. In most other electrical drawings, the signal paths are drawn as horizontal or vertical lines. It would be possible to draw a flip-flop in this manner, too, but most companies now use the angled lines.

The outputs of a flip-flop are such that if A is $+$, B is $-$, or vice versa. (We might call $A+$ *and* $B-$ the ON state and $A-$ *and* $B+$ the OFF state, for instance.) In the *T-type,* an input pulse at T will change the state of the circuit. If the flip-flop is initially ON a pulse will change it to OFF, and vice versa. Another type of flip-flop is the *set-reset,* which is shown by means of a block diagram in Fig. 6·6c. In this circuit, a pulse at the S (set) input causes the flip-flop to turn on or stay on, depending on its original state. A pulse at R (reset) causes the circuit to turn off, or stay off.

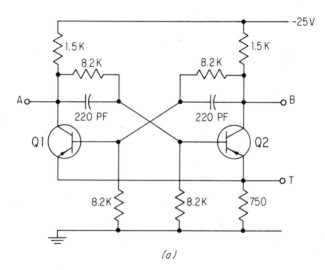

(a)

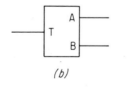 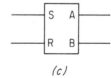

(b)   *(c)*

Fig. 6·6 A saturated flip-flop circuit. (a) Elementary diagram. (b) Block diagram of T-type and (c) S-R type. (From GE Transistor Manual.)

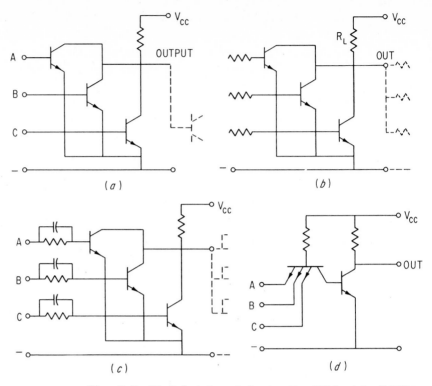

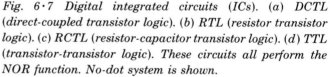

Fig. 6·7 Digital integrated circuits (ICs). (a) DCTL (direct-coupled transistor logic). (b) RTL (resistor transistor logic). (c) RCTL (resistor-capacitor transistor logic). (d) TTL (transistor-transistor logic). These circuits all perform the NOR function. No-dot system is shown.

Figure 6·7 shows several typical logic circuits that are manufactured as integrated circuits (explained in Chap. 7). Figure 6·7a is a direct-coupled circuit, not unsimilar to those of Fig. 6·4. The multiemitter transistor of the TTL circuit (Fig. 6·4d) can be economically fabricated, making the TTL circuit very adaptable to all forms of IC logic. [Note that this drawing uses the no-dot system for showing connections. (See Sec. 3·4.)]

6·6 Principles that apply to the preparation of elementary diagrams

1   Medium-weight lines are used for symbols and connecting leads or wires. Heavy lines may be used where emphasis of a component is desired.

2 A diagram should be arranged to give prominence to main features.

3 Uniform density is desired so that there are no congested areas and a minimum number of large white areas.

4 Very long lines or interconnecting leads are to be avoided as much as possible.

5 Lines are usually—and preferably—drawn vertically or horizontally, but they may be drawn in other directions when circumstances so dictate. Turns, or "bends," should be kept to a minimum.

6 A diagram should be arranged so that it reads functionally from left to right. In other words, the source, or input, is at the left, and the output is at the right. In complex drawings, the input is placed at the upper left, and the output at the lower right, if possible.

7 A complex drawing may be laid out in layers, with each layer reading from left to right or downward.

8 Connecting lines may be interrupted, where necessary, provided proper identification is added at the points of interruption.

9 A drawing should be arranged so that adequate space is available near each symbol for proper identification.

10 All lines should be routed as directly as possible, with a minimum number of zigzags, so that they cross a minimum number of other lines.

11 Components that are in series should look like they are in series; components that are in parallel should look like they are in parallel.

12 All component parts should be properly identified by means of reference designations.

A recent change in ANSI Y14.15 indicates that the no-dot system is preferred to the dot system for showing connections. Accordingly, we have shown several drawings (Figs. 6·7, 6·19, 6·21, 6·28, and 6·33) with this system. Because the author believes this system has weaknesses, he has shown most schematics with the dot system.

6·7 Reference designations

Symbols of all replaceable parts should be referenced. A reference may be placed above, below, or on either side of its part. ANSI-recommended practice includes giving each part a number, such as resistor $R5$, and its capacity, such as 100 ohms. Thus, we have two lines of designations (combinations of letters and numbers) for most elements in a circuit. Table 6·1 gives letters and examples that are typical of standard practice.

Table 6·1

element	letter	example
Capacitor	C	C5 10PF*
Inductor	L	L1 23MH*
Rectifier (metallic or crystal)	CR	$CR2$
Resistor	R	$R201$ 270
Transformer	T	T2
Transistor	Q	Q5 2N482 DETECTOR
Tube	V	$V3$ 6AU6 1ST I-F AMP

*MH stands for millihenry, a thousandth of a henry. PF stands for picofarad, a micromicrofarad (a millionth of a millionth of a farad). The term UUF is still sometimes used.

The letter X is sometimes used for transistors, although Q is recommended by both the Military and American National Standards. $R201$ does not mean the 201st resistor in the circuit. It means that it is the first resistor in the second (200 series) subassembly. Both the transistor and the tube contain a third line—their function. This is optional information insofar as the drawing is concerned.

6·8 Laying out an elementary diagram

Figure 6·8 shows a freehand sketch of a light-flasher circuit. This is the type of sketch that might be made by a design engineer or project chief. The purpose of the sketch is to furnish all the information necessary to make a finished drawing of this circuit and to construct it. If the sketch is very rough and there are many places where improvement is indicated, it may be desirable to make a new sketch.

A draftsman should produce a well-balanced work that is pleasing to the eye. In order to do this, he may have to change the configuration of the drawing (reorient some symbols and change some lines), but the circuit must still maintain its original technical significance. He must also strive for simplicity and clarity.

A glance at Fig. 6·8 reveals that, among other things, the transistors could be lined up better, it is a little crowded in the shaded region, and the sketcher has used loops for crossovers—a nonstandard procedure. Also the components do not have reference numbers.

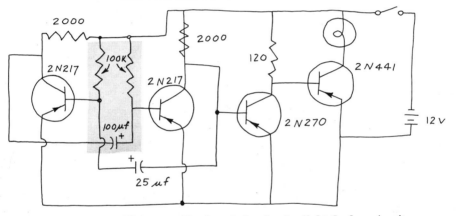

Fig. 6·8 *Freehand sketch of a light-flasher circuit.*

Figure 6·9 shows how one can go about correcting some of these deficiencies and starting the layout of the diagram. Figure 6·9a shows how the spacing between vertical lines in the crowded area can be determined by lettering the appropriate reference designations, possibly on a trial basis on a sheet of scratch paper, or on the new sketch if such is being made.

It is obvious that three, and probably all four, transistors can be aligned on one horizontal line to make for a more pleasing and professional-appearing drawing. Therefore a light construction center line can be drawn horizontally at about mid-height through the drawing area. At this time, or possibly later, after more components have been drawn, other

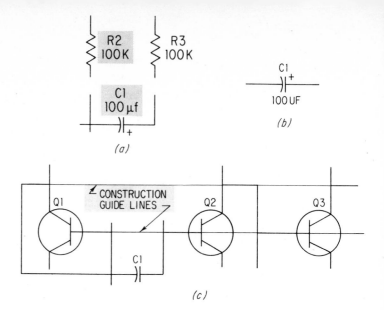

Fig. 6·9   Details in laying out an elementary diagram. (a) Allowing space for reference designations (identification). (b) alternate arrangement for referencing capacitor. (c) Using horizontal guidelines for transistors and other line work.

horizontal guidelines can be drawn in. One other such line is shown in Fig. 6·9c for the collector outputs.

The transistor envelope circles can be drawn in at this time if the spacing between them can conveniently be determined. Figure 6·9a has provided the spacing around $C1$, and by allowing a little more spacing on each side of the vertical lines (on which $R2$ and $R3$ are located), we can locate $Q1$ and $Q2$ without much fear of having to redraw the diagram. Because there are no components between the other transistors, we can locate them at this time without much difficulty. (Note that this is DC (direct-coupled) circuitry. If it were transformer-coupled, considerably more work and planning would be necessary.)

After putting in the bases, collectors, and emitters, one can draw more lines, such as appear in Fig. 6·9c, and be well on the way toward completing the circuit diagram. One logical step to perform next would be to draw resistor $R4$ or $R1$ (reoriented from the sketch).[1] Then all five resistors could

[1]The reader will have to look ahead to Fig. 6·11 to see which resistors have been assigned designations $R1$ and $R4$.

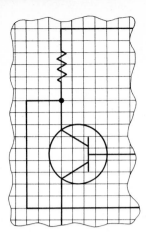

Fig. 6·10 Using a grid to lay out an elementary diagram. Using ⅛-in. (0.125) grid spacing (modules), transistor envelopes are often drawn five units, or ⅝ in., in diameter.

Fig. 6·11 The completed elementary diagram of a flasher circuit. (From a circuit in the RCA Transistor Manual.)

be located at the same level, and the uppermost horizontal line could be located.

Another popular way to lay out a circuit diagram is to use standard grids. Either a final full-size sketch can be made on sketching paper having grids, or an undergrid can be placed under the tracing paper or film on which the drawing is to be made. Figure 6·10 shows how the left end of the flasher circuit might be laid out, using this method. With 0.125 grid spacing, a ⅝-in. transistor envelope has been drawn. The collector-output line has been drawn too spaces away from the edges of the circle, and the resistor was started one and one-half spaces from the junction dot.

Figure 6·11 shows the finished elementary drawing of the flasher circuit. Standard ANSI referencing has been used. Transistor identification has been placed at the top of the diagram with slightly larger-than-average letters. The author has used this arrangement, rather than putting the designations near the transistor symbol, which, of course, is *also* approved practice and largely a matter of preference. Note that the entire drawing appears to be symmetrical and that a fairly uniform *density* is apparent. There are often notes of a general or specific nature at the bottom of an elementary diagram, and this one is no exception. Instead of locating transistor $Q4$ as shown we might have drawn a horizontal line from the output of $Q3$ to $Q4$, thus placing $Q4$ "above" the other transistors.

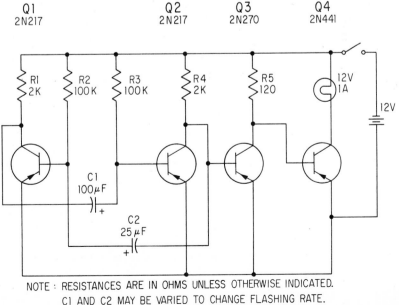

NOTE : RESISTANCES ARE IN OHMS UNLESS OTHERWISE INDICATED.
C1 AND C2 MAY BE VARIED TO CHANGE FLASHING RATE.

Some persons might prefer to put $Q4$ in this location because it is a typical DC circuit pattern and a zigzag would be eliminated.

6·9 Line arrangements according to voltages

Although the circuit of Figs. 6·9 to 6·11 does not become involved in this problem, some circuit diagrams will have lines representing several different voltages. A good practice is to draw the schematic diagrams so that the highest voltage is on the uppermost line, the next highest is on the line below, and finally, the most negative voltage is on the lowest line. This makes for easier reading of the elementary diagram in many cases. Typical voltage sequences might be as follows:

235	volts
150	volts
6.3	volts
0	volts
−45	volts

The top line is what is known as the $B+$ line; the zero line is the $B-$, or common, line and may be joined to the ground or chassis. Transistor circuits, as a rule, do not have as many lines as this, but tube circuits may have this many, depending upon the circuit and types of tubes.

6·10 Examples of elementary diagrams

Figure 6·12 shows the complete schematic diagram of a transistor radio. This particular circuit has a separate mixer and oscillator, two I-F stages, a detector and audio amplifier, and a class-B push-pull output. The transformer coils are drawn as a series of complete loops. This loop, or helix, is one of two approved symbols in ANSI Y32.2. However, Mil Std 15A approves only the more rudimentary symbol, such as is used in Figs. 6·17 and 6·19. Note that $L1$ has a metallic core, $L2$ has an air core, and $T1$ is adjustable (the line with an arrowhead). The author would be tempted to call $L1$ and $L2$ $T1$ and $T2$. (See Fig. 6.12.)

Fig. 6·12 Elementary (schematic) diagram of a transistor radio receiver. (From Milton S. Kiver, "Transistors," 3d ed., McGraw-Hill Book Company, New York, 1962.)

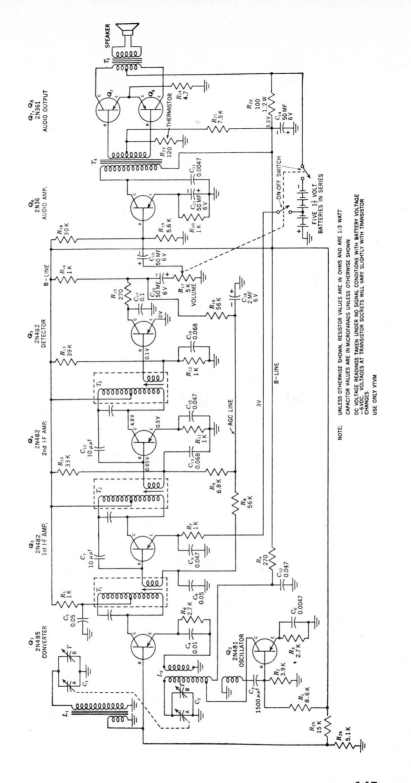

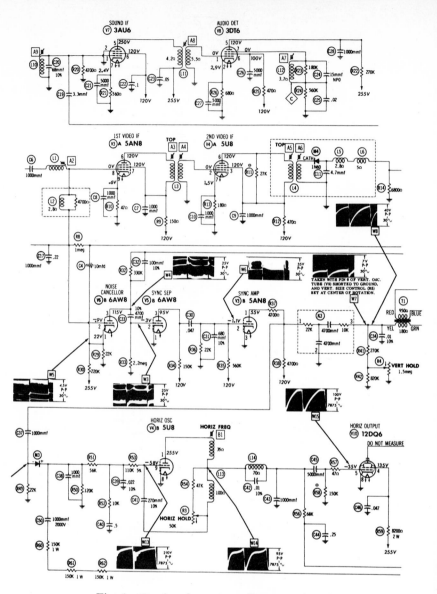

Fig. 6 · 13 An elementary diagram of a portable television receiver. (Courtesy of Howard W. Sams & Co., Inc.)

6 · 11 Tube circuits

Figure 6 · 13 shows a complete elementary diagram for a television receiver. The black rectangles with oddly shaped white lines in them show the waveform that the signal is supposed to have at the circuit location shown. This is very

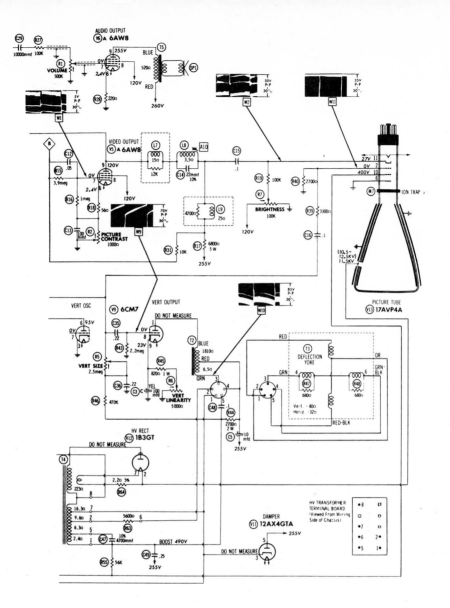

useful for testing and servicing circuits. (Figure 6·15 shows some typical waveforms.) The television diagram follows a pattern that is often used. At the top is the audio circuitry reading from left to right. Next down is the video I-F and output circuit terminating at the picture tube. Below this are the synchronization, vertical, and horizontal controls and the power supply. Aside from the video tube, this circuit utilizes the usual components found in many electronics circuits. The

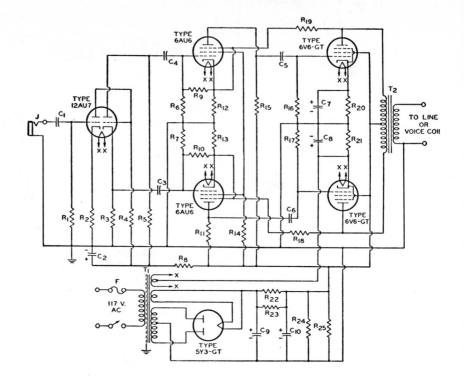

$C_1 = 0.1\ \mu\text{f, paper, 600 volts}$
$C_2 = 40\ \mu\text{f, electrolytic, 450 volts}$
$C_3,\ C_4 = 0.02\ \mu\text{f, paper, 600 volts}$
$C_5,\ C_6 = 0.05\ \mu\text{f, paper, 600 volts}$
$C_7,\ C_8 = 50\ \mu\text{f, electrolytic, 50 volts}$
$C_9,\ C_{10} = 80\ \mu\text{f, electrolytic, 450}$
volts
$F = \text{Fuse, 1 amp}$
$R_1 = 470{,}000\text{ ohms, 0.5 watt}$
$R_2 = 6{,}800\text{ ohms, 0.5 watt}$
$R_3,\ R_5 = 39{,}000\text{ ohms } \pm 1\text{ per cent,}$
matched, 1 watt
$R_4 = 220{,}000\text{ ohms, 0.5 watt}$
$R_6,\ R_7,\ R_{14} = 1\text{ meg, 0.5 watt}$
$R_8 = 10{,}000\text{ ohms, 1 watt}$

$R_9,\ R_{10},\ R_{11},\ R_{15},\ R_{16},\ R_{17} = 330{,}000$
ohms, 0.5 watt
$R_{12},\ R_{13} = 1{,}800\text{ ohms } \pm 1\text{ per cent,}$
matched, 0.5 watt
$R_{18},\ R_{19} = \text{carbon-film type,}$
100,000 ohms ± 1 per cent,
matched, 2 watts
$R_{20},\ R_{21} = 510\text{ ohms, 2 watts}$
$R_{22},\ R_{23} = 390\text{ ohms, 2 watts}$
$R_{24},\ R_{25} = 150{,}000\text{ ohms, 2 watts}$
$T_1 = \text{power transformer,}$
350-0-350 volts rms, 125 ma
$T_2 = \text{output transformer for}$
matching line or voice coil impedance to 9,000–10,000-ohm plate-
to-plate tube load

Fig. 6·14 *A schematic diagram of a high-fidelity audio amplifier. (From RCA Receiving Tube Manual.)*

company that made the drawing does use, however, circles and squares to enclose component identifications, which because of their comparatively large numbers do give the diagram a distinctive appearance. Voltages of conductor paths are indicated at many points throughout the drawing.

Figure 6·14 shows the drawing of a high-fidelity audio amplifier circuit. This drawing differs from those discussed so far in that the values are placed in a table below the figure. Such an arrangement may make for slower reading of the diagram but has the advantage of supplying more information about each part than can be conveniently shown with the other system. This drawing would be read as follows:

1 The audio signal (single-path, single signal) enters at the input jack and almost immediately enters the amplifier part (left half) of tube 12AU7. This tube is both an amplifier and phase inverter twin triode.

2 The amplified signal then enters the grid of the second half of the same tube. At this point two separate outputs are developed which are 180° out of phase with each other, but otherwise equal.

3 One output goes through capacitor $C4$, and is used to supply one of the 6AU6 pentodes, which are audio amplifiers. The other output goes through $C3$ and supplies the other 6AU6. Both 6AU6 tubes are connected as straight amplifiers.

4 At this stage there are two signals 180° out of phase, each of which is fed to one of the 6V6-GT tubes which are connected in push-pull.

5 After leaving the push-pull amplifiers, the two signals are combined in the output transformer $T2$ to produce one output signal.

6 Resistors $R9$, $R10$, $R18$, and $R19$ form a feedback network that straightens out the nonlinearity of the amplifier.

6·12 Commonly used units for components

In addition to units such as henrys and ohms, certain prefixes (multipliers) are used when components have extremely large or small values. The most-often-used prefixes are shown in Table 6·2. Suggested units with their prefixes appear in Table 6·3.

Some lettering can be saved by using notes such as the

one shown in Fig. 6·11 about resistance values. A similar note might be: *All capacitance values are in microfarads unless otherwise shown.*

Another helpful device is that shown in Table 6·4, which shows the highest number of each component, as well as omissions, in a diagram.

The reader may wonder why two resistors were not used. There could be several reasons for such an occurrence. One possibility would be that the original design was altered after testing or use and a particular component was taken out. If, later, another one is added, it is usually given a new number, not the number of the deleted one.

Table 6·2

prefix	multiplier	symbol
Mega	10^6	MEG
Kilo	10^3	K
Milli	10^{-3}	M or MILLI
Micro	10^{-6}	μ or U
Pico	10^{-12}	P (or $\mu\mu$ or UU)

6·13 Waveforms

Some common waveform symbols are shown in Fig. 6·15. These are sometimes shown on an elementary diagram, as was done in Fig. 6·13. Some common forms that are not shown are sawtooth, trapezoidal, rectangular, exponential, clipped, and triangular.

6·14 Mechanical linkage and other mechanical arrangements

In Fig. 6·12, at the left side, are two capacitors connected by means of dashed lines. The dashed (dotted) lines mean that the two devices are connected mechanically, so that when one is turned, the other is turned to the same position simultaneously. Figure 6·16 shows two other situations in which mechanical connections are shown. Figure 6·16a is part of a diagram that is part pictorial, part schematic, and part con-

Table 6·3

range	units to be used	examples
Up to 999 ohms	Ohms	520 or 210Ω
1,000–99,999 ohms	Ohms or kilohms	45K or 45,000
100,000–999,999 ohms	Kilohms or megohms	500K or 0.5 MEG
10^6 or more ohms	Megohms	2.5 MEG
Up to 9,999 pico-farads (micromicrofarads)	Picofarads	150 PF (old: 150 UUF)
10,000 or more picofarads	Microfarads	0.05 UF (or 0.05 MFD)
Up to 0.001 henry	Microhenrys	5 UH (= 0.000005 henry)
0.001–0.099 henry	Millihenrys	3 MH (= 0.003 henry)
0.1 or more henry	Henrys	20H

nection. The mechanical path may be easily traced. The auto-positioner works as follows:

1 The operator turns a remote-control switch to a certain position.
2 The relay (slow-operating in this example) is energized, lifts the pawl out of the sprocket wheel, and also turns on the motor.
3 The motor drives the autopositioner shaft until the position corresponding to the new position of the remote control is reached.
4 The relay circuit now opens, the pawl engages the stop wheel, and the energy to the motor is cut off.

Ganged tuning capacitors (Fig. 6·16b), like ganged switches, may be shown on an elementary diagram. Figure

Table 6·4

HIGHEST REFERENCE DESIGNATIONS
C23   L7   R43
NOT USED
R22, R23 C16

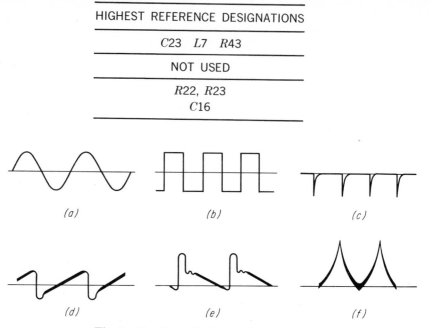

(a) (b) (c)

(d) (e) (f)

Fig. 6·15   Symbols of waveforms. (a) Sine wave. (b) Square wave. (c) Trigger. (d), (e), and (f) Complex forms.

6·16c shows mechanical connections between the attitude indicator and synchro of a DC-10.

6·15 Separation and interruption

It is often desirable to show a separate package apart from the rest of the circuit or system. For instance, if the same package (or *pack* as it is sometimes called) is to be used several times, it may be drawn as blocks instead of as an elementary diagram. In such a scheme, the elementary would have to be drawn once, but only once. In Fig. 6·17 three ways of drawing a pack schematically are shown using different terminal or interruption techniques. Figure 6·17d shows how the pack might be drawn diagrammatically as a block. Figure 6·17a shows the circuit as removed at terminals. Figure 6·17b simply encloses the pack with an optional enclosure line but does nothing about the terminals. (In all four figures, the lines, or terminals, have been identified by means of numbers.) Figure 6·17c shows another method of separating a

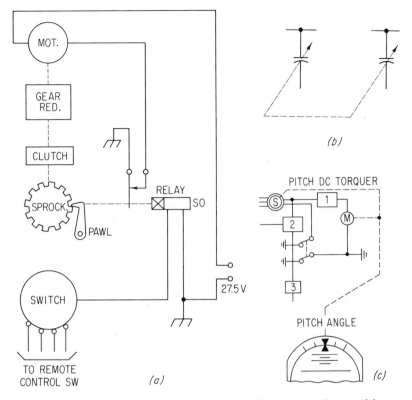

Fig. 6·16 Mechanical connections. (a) Autopositioner. (b) Granged tuning capacitors. (c) Attitude indicators.

package. While this is a fairly widely used method, it is slightly confusing. The V-shaped lines at the end of each conductor do not represent any specific kind of connection in this scheme. Yet ANSI Y32.2 uses this symbol to represent a male contact. It is so used in other drawings in this book, especially in Chap. 9. The V's, incidentally, are not supposed to be arrowheads. They are correctly drawn at a 90° angle.

Figure 6·18 illustrates other ways of interrupting groups of lines. Lines are generally grouped together in threes. If there are more than three lines to be shown, groups should be separated, as at the right in Fig. 6·18a. Spacing between lines should be a minimum of ¼ in., and between groups ½ in. Brackets may be joined by means of dashed lines, as in Fig. 6·18b. Figure 6·18c shows hypothetical uses of circuit-return symbols. If the inverted triangle is used, letters and general notes (Fig. 6·18d) will probably be necessary.

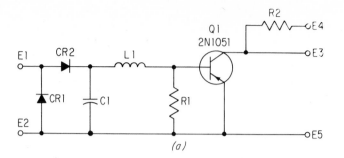

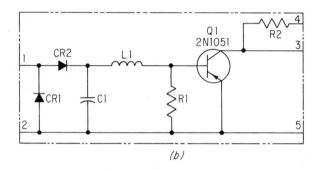

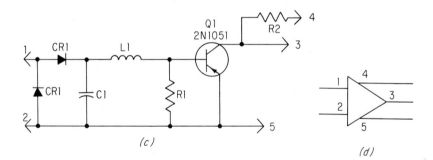

Fig. 6·17 *Representation of a circuit package separated from the rest of the system.* (*From ANSI Y14.15, "Electrical and Electronic Diagrams."*)

The chassis symbol and the first triangle (with letter F) serve the same purpose, and either one could be used. These symbols do not have to be oriented as shown. If it is more convenient, they can be drawn at the ends of lines extending upward or to the right or left. In other words, the entire figure (Fig. 6·18c) could be turned at 90 or 180° from its present position. (Letters and numbers might be changed to read in a convenient manner, though.)

6·16 Other examples of schematic diagrams

Figure 6·19 shows the elementary diagram for the preamplifier circuit of the Apollo moon camera. It is enclosed by the optional enclosure lines. The preamplifier is made up of discrete components, primarily to provide low-noise performance. The video signal from the SEC camera is fed into the preamplifier. Its input is a field-effect transistor ($Q1$) stage with a 330-k or kΩ load resistor. This is followed by a feedback pair.

Other parts of the camera are a post amplifier, deflection circuits, and a power supply (two, in fact—high and low voltage). In addition to the camera there is a small (85-cc) viewfinder monitor for the astronaut and a transmitter. Figure 5·2

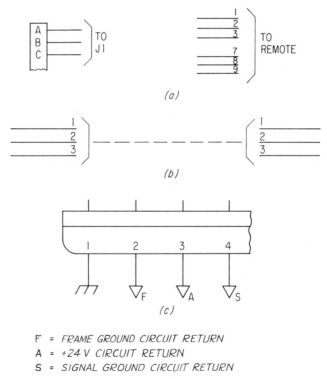

(a)

(b)

(c)

F = FRAME GROUND CIRCUIT RETURN
A = +24 V CIRCUIT RETURN
S = SIGNAL GROUND CIRCUIT RETURN

(d)

Fig. 6·18 Interruption and circuit return. (From ANSI Y14.15.)

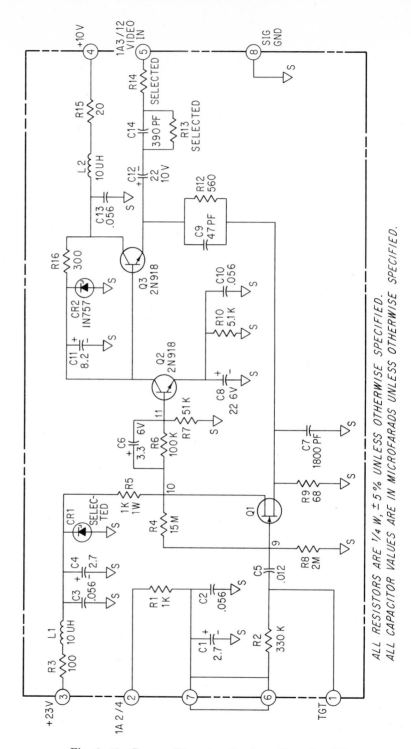

Fig. 6·19 Preamplifier circuit of Apollo color TV camera.
(Westinghouse Electric Corp.)

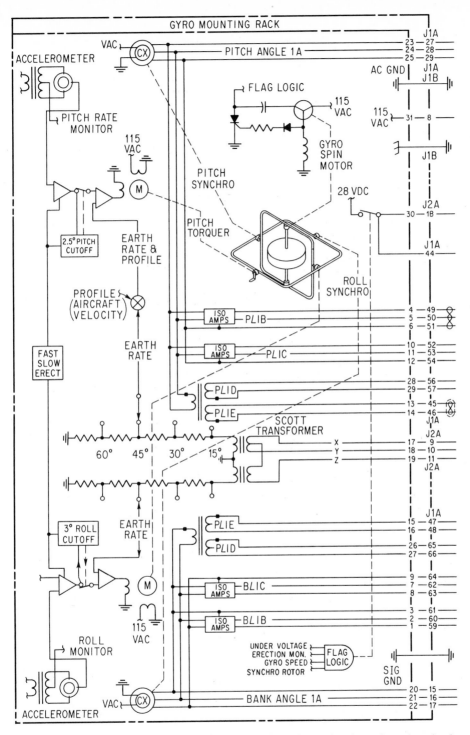

Fig. 6·20 Part of a systems schematic of an aircraft attitude system. This is a hybrid drawing. (Douglas Aircraft Company.)

shows the entire system in block form. The no-dot system has been used.[2]

The camera is 17 in. long, including a zoom lens, and weighs 13 lb. It generates a field sequential color signal using a single-image tube and a rotating filter wheel. A ground-station color converter later changes the sequential color signal to a standard NTSC color signal. The lens has a $.T$ number of 5.1 to 51, zoom ratio of 6:1, and focal length of 25 to 150 mm.

Figure 6·20 shows a "system schematic" drawing (about the left third) produced by a major aircraft manufacturer. This is a *hybrid* diagram that combines information from wiring and elementary and logic diagrams as well as a little pictorial drawing. Other systems schematics may include hydraulic systems and mechanism-type drawings as well as information of the type shown in Fig. 6·20. These hybrid diagrams are of extreme value to persons engaged in maintenance and trouble-shooting operations.

The reader will notice a loop-like symbol at the very right edge of lines 49 to 51. This indicates that the three wires are twisted together. Somewhat below this is a similar symbol which indicates that wires 45 and 46 are twisted together and shielded.

SUMMARY The schematic or elementary diagram shows by means of graphic symbols the functions and connections of a specific circuit arrangement. Symbols for use in such a diagram are covered in a Military Standard or an American National Standard, and the preparation of the diagram itself is covered by ANSI Y14.15. There are certain basic arrangements of transistors and tubes which are usually repeated in electronics circuit diagrams. With these and some types of interstage coupling, certain patterns are often discernable—usually early in the game, when one starts to plan an elementary diagram. Designation, or referencing of each component part of a circuit, is important. Certain standard abbreviations and prefixes are used for this purpose. Sufficient space must be made available near each component for referencing.

Conventional treatment is sometimes employed to eliminate the drawing of certain lines in order not to have too cluttered a drawing. On the other hand, additional material, such as general notes and data on waveforms, is often added

[2] There are connections at points 9, 10, and 11.

to a schematic diagram. A symbol is generally spaced midway along the span of circuit path on which it is drawn. Mechanical connections must be shown at times. Certain patterns are usually followed in laying out certain types of circuits in diagrammatic form. Radio circuits, for example, usually follow a *cascade* arrangement in which the signal path goes from left to right and the tubes are aligned horizontally. Auxiliary circuits, such as power circuits, are generally placed in the lower part of any schematic diagram. Overall balance and symmetry are other concepts that are observed in diagrammatic layout.

QUESTIONS

6·1   What set of standards governs the preparation of elementary diagrams?

6·2   In what way, or ways, is a schematic diagram different from a connection diagram?

6·3   How would you show the value 100,000 ohms on a circuit drawing? 150,000 picofarads?

6·4   To what sort of elements do the following terms refer:

   a. 12AV6   b. *L2*   c. 50K
   d. 2N329   e. *CR2*

6·5   What is meant by the term *density* as it applies to elementary diagrams?

6·6   What is meant by the term *symmetry* as it applies to elementary diagrams?

6·7   Does an elementary diagram sometimes accompany another type of electrical diagram? If so, what type?

6·8   For what purposes are schematic diagrams used?

6·9   Where are auxiliary circuits normally placed on the elementary diagram?

6·10   If a resistor is positioned on a vertical line (signal path), where would you place its identification?

6·11   Where would you place the identification for a vacuum tube in a schematic diagram of a circuit having cascade projection?

6·12   In reading an elementary diagram of a multistage circuit, how would you go about determining whether the tube heaters were in series or parallel?

6·13   What elements are often found in a hybrid diagram?

6·14   What do the parallel lines adjacent to the outline of a picture-tube symbol represent?

6·15 Define the prefixes micro-, milli-, kilo-, and mega-.

6·16 Is it possible to draw a schematic diagram using only one width of line and in so doing comply with the standards?

6·17 If you are assigned to make a finished drawing of an elementary diagram of a circuit, in what form is the information to which you will refer apt to be?

6·18 What are three basic functions which are performed by transistors?

6·19 What methods are used in coupling different stages together? (Name three.)

6·20 In what way may the number of connecting lines in a diagram be reduced?

6·21 How many inputs may a T-type flip-flop circuit have?

6·22 State briefly, or use sketches to show, how you would separate a package from the rest of a circuit.

6·23 What is the difference between a picofarad and a micromicrofarad?

6·24 If an SR-type flip-flop circuit is in the OFF position, what effect does a pulse at R have?

PROBLEMS 6·1 Draw the elementary diagram of the circuit shown in Fig. 6·21. Use standard ANSI identification for each component, adding value or catalog number as follows:

$CR1$	1N914	$R6$	1.5K
$CR2$	1N914	$R7$	1K
$R1$	1000 ohms	$R8$	510 ohms
$R2$	51K	$R9$	5.1K
$R3$	5.1K	$R10$	1000 ohms
$R4$	510 ohms	$C1$	20V 2.2 UF
$R5$	51K	S	ground

If you think the numbering sequence of any or all devices can be improved, do so. Add a suitable title and the following note: Unless otherwise specified all resistors are ±5%, ¼ watt. There is a connection at pt. 11 but none at pt. 10. Use 8½ × 11 paper.

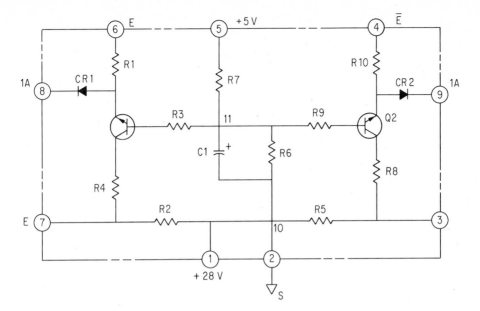

Fig. 6·21 (Prob. 6·1) Elementary diagram of filter circuit of Apollo TV camera. (No-dot system shown here.)

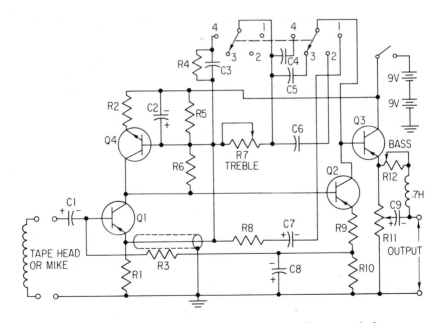

Fig. 6·22 (Prob. 6·2) Elementary diagram of phono-tape preamplifier.

6·2 Draw the preamplifier circuit shown in Fig. 6·22 as a complete elementary diagram. The drawing as shown is slightly crowded in the right one-third. Correct this deficiency. Use ANSI practice in identifying and giving information about the components. Reletter or renumber any components which you think are not identified in the proper sequence. The following information is pertinent:

$Q1$	2N508	$R9$	100 ohms
$Q2$	2N508	$R10$	1000 ohms
$Q3$	2N322	$R11$	5000 ohms
$Q4$	2N634A	$R12$	50K
$R1$	47 ohms	$C1$	20 UF 20V
$R2$	100 ohms	$C2$	50 UF 3V
$R3$	82K	$C3, C5$	.01 UF
$R4$	15K	$C4$	.04 UF $\pm$10%
$R5$	15K	$C6$	.005 UF
$R6$	820K	$C7$	5 UF
$R7$	30K	$C8$	100 UF
$R8$	7.5K	$C9$	10 UF 15V

Show an appropriate title. Use 11 × 17 or 12 × 18 paper.

6·3 Complete the elementary diagram of Fig. 6·23 by insertion of the correct symbols as indicated by standard notation. Add the identification of each component and the capacity if given below. Squares $Z1$, $Z2$, etc., are integrated circuits drawn as squares with manufacturer's identification shown within the square. Video input is at 12.

$Z1$	SE501G	$C10$	15 PF
$Z2$	MD2905F	$R1$	20K
$Z3$	SE501G	$R2$	510
$Z4$	SE501G	$R5$	51K
$CR1$	IN914	$R6$	10
$CR2$	IN914	$R7$	470
S	ground	$C11$	6800 PF
$C1$	3.3 UF	$R4$	3.3 K
$C2$	3.3 UF	$L1$	10 UH
$C5$	22 UF	$L2$	10 UH
$C6.$	.056 UF	$C14$	3.3 UF
$C7$	22 UF		
$C8$	056 UF		
$C9$	15 PF		

This drawing will be slightly crowded on 8½ × 11 paper, and will have more than enough room on 11 × 17 paper.

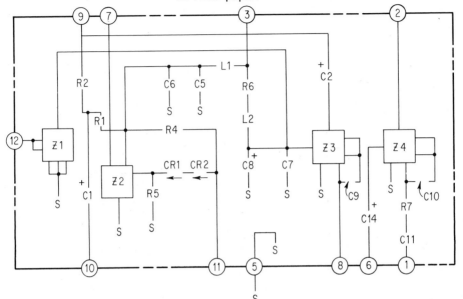

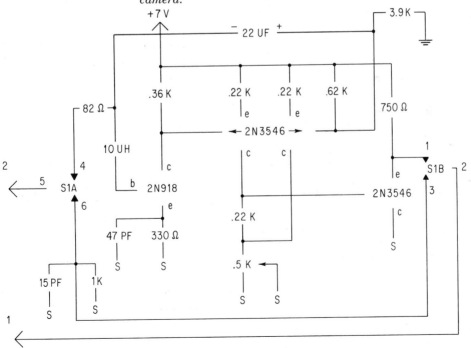

Fig. 6·23 (Prob. 6·3) Postamplifier circuit of moonwalk TV camera.

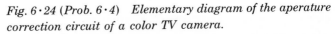

Fig. 6·24 (Prob. 6·4) Elementary diagram of the aperature correction circuit of a color TV camera.

6·4   Complete the elementary diagram of the aperture correction circuit of the color TV camera shown in Fig. 6·24. Transistor configurations are represented by the letters c and e, representing collector and emitter leads. Transistors are PNP type, except 2N918 which is NPN. Other components are indicated by their quantities. Add the note: Except as otherwise noted all transistors are ¼ watt. Use 8½ × 11 paper.

6·5   Complete the elementary diagram of the FM tuner circuit shown in Fig. 6·25. $C1$ and $C6$ are ganged, 10-365 PF; $C3$ and $C4$ are .01 UF; all other capacitors are 10 PF, except $C8$ and $C10$ which are 250 UF. $R1$ is 180 ohms; $R2$, 5K; $R3$, 33K; $R4$ and $R6$, 1K; $R5$, 10K; $R7$, 150K; $R8$, 1.5K; $R9$, 470K; $R10$, 7000. Use 11 × 17 or 12 × 18 paper.

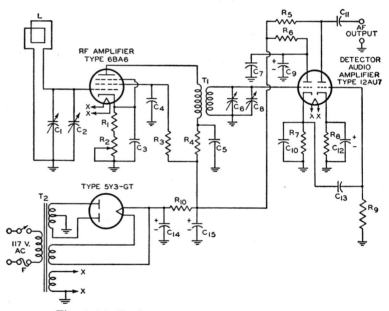

Fig. 6·25 (Prob. 6·5) Elementary diagram of FM tuner circuit.

6·6   Redraw the high-fidelity amplifier shown in Fig. 6·14 to about 2½ or 3 times the size that it appears in the book. Instead of the component identification system shown, use the ANSI system of identification shown in this chapter. Use 11 × 17 or 12 × 18 paper.

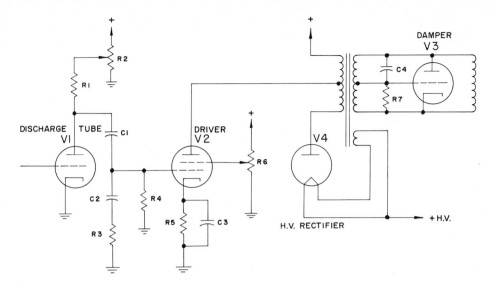

Fig. 6·26 (Prob. 6·7) Elementary diagram of scan-generator circuit.

6·7 Redraw the schematic diagram of the scan-generator circuit to approximately twice the size it is shown in Fig. 6·26. Identify each component as $C1$, $V1$, $R2$, etc. Use 8½ × 11 paper.

6·8 Draw the audio portion, only, of the television-receiver circuit shown in Fig. 6·13. Make the tube symbols about 1 in. in diameter and draw the other symbols to size, proportionately. Use 11 × 17 or 12 × 18 paper.

6·9 Make a complete elementary diagram of the sweep-fail-protect circuit shown in Fig. 6·27. Some of the symbols are missing but are identified by their values. Some symbols shown are incorrect. $Q2$ is a double transistor, and $Q1A$ is a double transistor of the same type that is separated. Collector and emitter connections are indicated by the letters c and e, respectively. $Z1$ is an integrated circuit which should appear as shown in Fig. 6·27. One line is missing, but its location is fairly obvious. Your completed drawing should have the correctly drawn and oriented symbols and appropriate standard reference designation for every symbol. The following information, when put on the diagram, will complete the schematic drawing: All

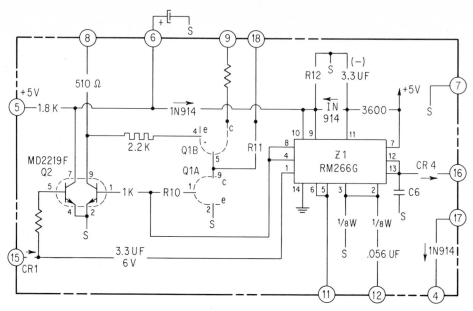

Fig. 6·27 (Prob. 6·9) Diagram of a sweep-fail-protect circuit. (Westinghouse Electric Corp.)

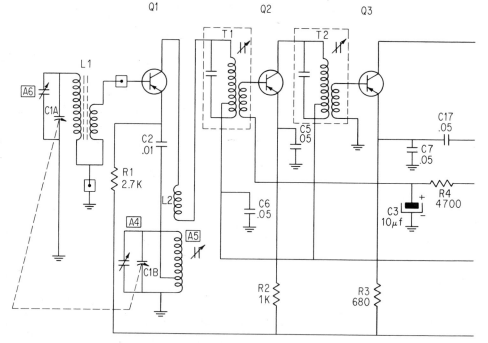

Fig. 6·28 (Prob. 6·10) Elementary diagram of transistor clock radio.

resistance values are in ohms unless otherwise indicated and are ¼ watt. All capacitance values are in microfarads unless otherwise indicated. Use 11 × 17 or 12 × 18 paper.

6·10 Make a drawing of the clock-radio circuit shown in Fig. 6·28. Correct the symbols where obsolete or incorrect ones appear. Add identification for semiconductors as follows: $Q1$, converter, 95101; $Q2$, first I-F, 95103; $Q3$, second I-F, 95102; $CR1$, detector, IN295; $Q4$, audio driver, 95201; $Q5$ and $Q6$, audio outputs, 95220; $CR2$, rectifier, IN290. All resistors are ½ watt. The no-dot system has been used except at connections $J1$ and $J2$. Resistance values are in ohms unless otherwise noted, and all capacitance values are in microfarads. Use 11 × 17 or 12 × 18 paper.

6·11 Figure 6·29 is a sketch of a volt-ohmmeter circuit with many of its symbols (resistors and capacitors)

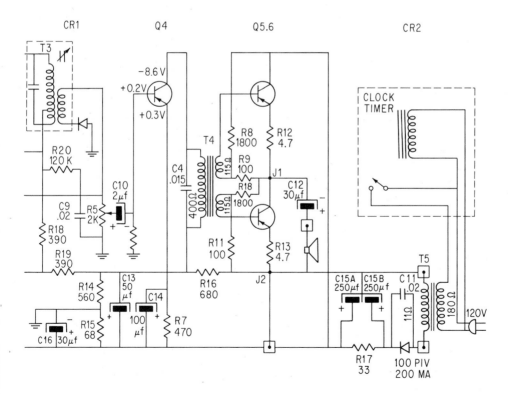

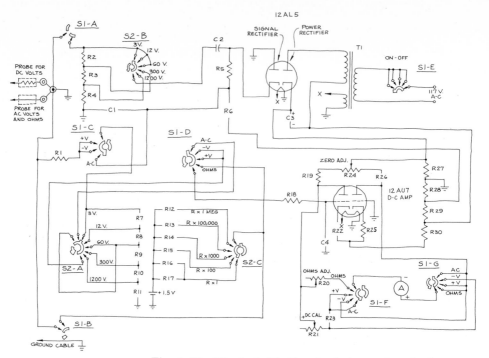

Fig. 6·29 (Prob. 6·11) Sketch of electronic volt-ohmmeter circuit. (From RCA Tube Manual.)

missing. Make a complete elementary drawing of this circuit. Values are as follows:

C1	0.1 UF	R13, 18	1 MEG
C2	0.33 UF	R14	10K
C3	10 UF	R15	1000 ohms
C4	0.01 UF	R16	10 ohms
R1	5 MEG	R17	330 ohms
R2	800K	R19	15K
R3	1.36 MEG	R20	15K
R4	250K	R21	7.5K
R5	67.8K	R22, 25	1.5K
R6	36.1	R23	470 ohms
R7	3.7 MEG	R24	12.5K
R8	1 MEG	R26	12K
R9	.2 MEG	R27	47K
R10	37.5K	R28	130K
R11	12.5K	R29, 30	68K
R12	10 MEG		

Use 11 × 17 or 12 × 18 paper.

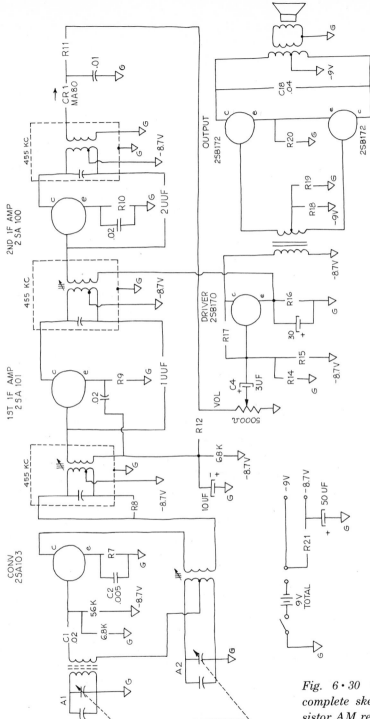

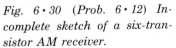

Fig. 6·30 (Prob. 6·12) Incomplete sketch of a six-transistor AM receiver.

6·12   Figure 6·30 is a sketch of a six-transistor radio-receiver circuit. Many symbols are missing, but clues to their identity are given as reference designations or values. All transistors are PNP; their collector and emitter leads are identified. Capacitors at $A1$ and $A2$ are variable; no values are available. Capacitors in 455 KC packages also have no values. Other capacitances are given as microfarads (UF) unless otherwise shown. Resistance values not already shown are as follows:

$R7$	2.2K	$R15, R18$	33K
$R8, R16$	100K	$R17$	220K
$R9, R10$	1K	$R19$	680 ohms
$R11$	470 ohms	$R20$	5 ohms
$R12$	3900 ohms	$R21$	100 ohms
$R14$	5600 ohms		

Some symbols are incorrectly shown. Reference designations may not be in the best sequence. Correct and complete the elementary diagram of this circuit, using the standard symbols and reference-designation procedure. Use 12 × 18 paper or larger.

6·13   Figure 6·31a shows the schematic diagram of an integrated circuit that uses TTL logic wherein a multi-emitter transistor replaces input diodes or single-input transistors used in other similar circuits. This NOR gate can be converted to an AND gate by replacing the area marked (c) with the partial circuit shown at (b). Redraw the circuit shown in Fig. 6·31a, substituting the partial circuit (b) in the enclosure (c). Resistors may be drawn with rectangles or other approved symbols. Resistors should be numbered and the transistors should be renumbered for good sequential numbering procedure. Vcc utilizes pin #4, and zero-pin #6. Use $8\frac{1}{2}$ × 11 or 11 × 17 paper.

6·14   Figure 6·32 is a sketch of the logic diagram and elementary diagram of a flip-flop circuit. Make a complete instrument drawing of this sketch, improving symbology or layout wherever you think it is advisable. Add the following information: total capacitance: 115 PF; total resistance: 70K; tunnels (crossover paths): 14; Vcc: Pin #3; Gnd: Pin #8.

Use 11 × 17 or 12 × 18 paper.

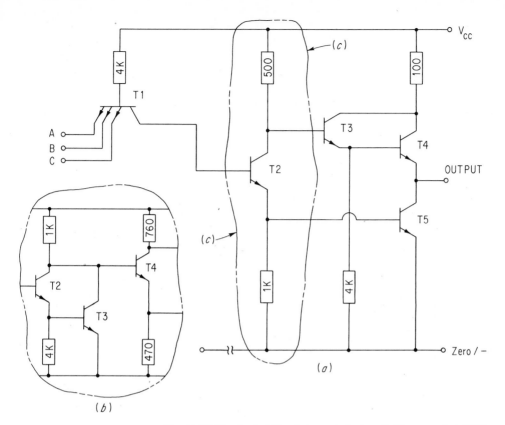

Fig. 6·31 (Prob. 6·13)   Integrated-circuit diagram. (a) TTL NOR-gate circuit. (b) Partial circuit for substitution.

6·15   Figure 6·33 shows an incomplete sketch of a satellite beacon oscillator. Many symbols are missing or incorrect, but clues are present as to their correct identities and/or values. All transistors are PNP; their emitter and collector leads are identified. Capacitors are identified as $C1$, $C2$, or with symbols, some obsolete or nonstandard. Complete the elementary diagram insofar as possible using standard symbols and reference designations.

Some values are: $R1$, 1780 ohms; $R2$, 5210 ohms; $R5$, 178; $C1$, $C4$, $C9$, and $C12$, 1 UF; $C3$, 0.1 UF; $C5$, $C8$, and $C11$, 0.01 UF; $C2$ and $C6$, 90 UF; and

C7, C10, and C13, 1000 UF. L1, L2, and L3 are F54796. *TR*1 is F54871; *TR*2 is F54870; *TR*3 is F54868; and *CR*1 is F54857.

Use 11 × 17 or 12 × 18 paper.

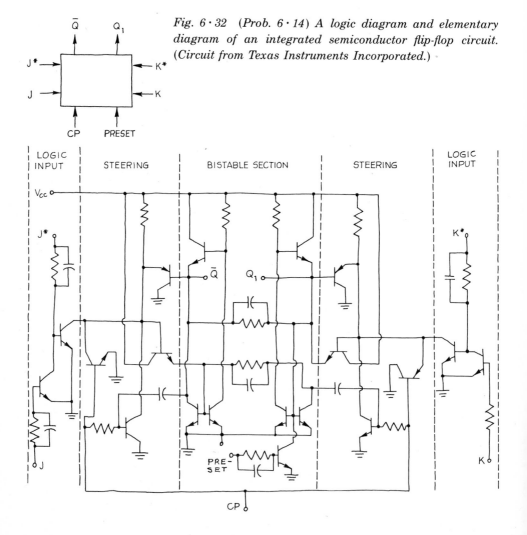

Fig. 6·32 (Prob. 6·14) A logic diagram and elementary diagram of an integrated semiconductor flip-flop circuit. (Circuit from Texas Instruments Incorporated.)

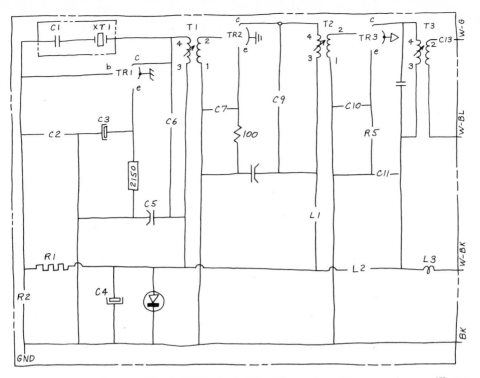

Fig. 6·33 (Prob. 6·15) Elementary diagram of an oscillator for a communications satellite. Rough sketch.

Chapter 7
Miniaturization
and Microelectronics

Nowhere but in the electronics field has miniaturization become so evident to the interested or even the casual observer. This concept, which is still evolving, may be divided into three broad classes:

1 Discrete-component-parts approach
2 Integrated-circuit approach
3 Hybrid circuitry

In this chapter, the author will show and explain some typical drawings and the graphical approach to certain problems in miniaturization. The first section will deal with the discrete-component approach.

Discrete-component-parts approach

7·1 Printed circuits

Two developments gave major impetus to miniaturization. One was the invention of the transistor, and the other was the development of printed circuitry. The combination of the two has brought great advances to the civilian consumer-product area, the military effort, and the space program.

Figure 7·1 shows photographs of two types of circuits, the printed-board type (in this case part of a package for an analog computer), and the thin, flexible laminate type. A set of drawings for a plug-in board will be shown following an explanation of certain requirements and procedures.

(a)

(b)

Fig. 7·1 Printed circuits. (a) Printed board as part of a plug-in unit. (b) A flexible laminate board. (Electronics Associates, Inc., and Sandia Corp.)

The great bulk of printed circuits are made from specially prepared boards (such as laminated paper phenolic or lucite) which are coated on one or both sides with a thin layer of conducting material called a *foil*. The boards are usually of standard thicknesses from ⅟₃₂ to ¼ in., and the foil, usually of copper, aluminum, copper-clad aluminum, or brass, is usually of one of three standard thicknesses:

0.00135 in., called *1 ounce*
0.0027 in., called *2 ounce*
0.0040 in., called *3 ounce*

From a master drawing, the desired pattern of conductor paths may be put directly on the board by one of three methods: (1) offset printing, (2) photoengraving, or (3) silk screen. This puts a pattern of acid-resistant coating on the foil surface which, when the board is immersed in an etchant solution, remains while the rest of the metal is eaten or etched away by the solution. Printed circuits are sometimes referred to as *etched circuits* because of this particular step in their production.

Then holes are drilled, and components are mounted (usually on the opposite side from the printed circuitry if the pattern is placed on only one side) and are soldered in place (generally by dipping the wiring side into a solder bath). A protective coating may be put on the board and components.

For some elaborate circuits, the following drawings may be needed:

1 Final layout of components and conductor paths
2 Printed wiring master layout (made from the above drawing)
3 Drilling drawing (made from the final layout)
4 Marking drawing, including board number, special instructions
5 Assembly drawings
   a. For component installation
   b. Mechanical assembly of board to frame, etc.
   c. Wiring, if any

For simpler devices or circuits, three drawings may suffice.

7·2 Requirements for laying out a printed circuit and making drawings

Certain design requirements require careful consideration both before and during the making of the drawings for a printed circuit. Some of these are (1) size of board available, (2) type of connectors, (3) through connections such as eyelets, (4) shielding requirements, (5) location of crossovers, (6) maximum current, and (7) peak potential differences.

For example, spacing between conductor paths depends on the peak voltage potential. Table 7·1 shows suggested spacing.

Table 7·1

minimum conductor spacing	potential difference
0.031 in.	0–150V
0.062 in.	151–300V
0.125 in.	301–500V

The cross section of the conductor path should be large enough to carry the required current with a temperature rise of no more than 40°C. Table 7·2 is a partial table that provides some of this information.

The complete table appears in Appendix B. A more-or-less standard width of conductor path is 0.062 ($\frac{1}{16}$) in., and minimum spacing between conductors is the same. However,

Table 7·2

current (amperes)	0.00135-inch copper	0.0027-inch copper
1.5	0.015 wide	
2.5	0.031 wide	0.015 wide
3.5	0.062 wide	0.031 wide

heater circuits may require more width, and some paths have been made as small as 0.010 in. wide in special circumstances.

For these or similar reasons, *pads* (also called *lands* or *donuts*) around connections have to be of specified sizes. In exacting cases, sharp corners should be avoided by rounding the paths at turns and by gradually enlarging circuit paths as they approach connection pads. Obviously, *crossovers* should be kept to a minimum. This is why there are so many peculiar patterns among printed circuits. They are designed to avoid having circuit paths cross. However, it is possible to make an insulated (or bare) jumper (mounted like a component, such as a resistor), if a crossover is unavoidable.

It might also be well to mention that areas or lines of copper are sometimes left on the pattern for shielding or to take the place of chassis connections. A minimum distance between an outside foil path and the edge of the board is a must (0.031 in. is a common figure). Standard grids are often used, and connections from one side of the board to another can be made by one of several means. Figure 7·2 gives some of this information.

7·3 Making drawings for printed circuit

The first step in making a set of drawings for a printed circuit is to study the elementary diagram. Such a study will provide such information as:

1 What groups of components have common connections
2 What the peak potential differences may be
3 Grounds, heater lines, B + lines, etc., that are required
4 Other special pieces of information

The next step is to make a *preliminary sketch* (or maybe several) in which all the components are roughly located and

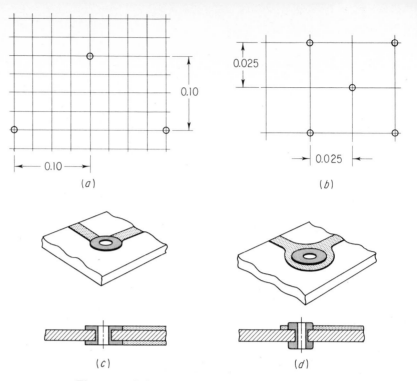

Fig. 7·2 Grid systems and through connections. (a) 0.1-in. grid. (b) 0.025-in. grid. (c) Plated hole connection. (d) Plated eyelet. (The 0.05-in. grid is not shown.)

circuit paths sketched in. (This is in reality a rough wiring diagram.) Now this sketch can be studied to see if:

1 Crossovers are at a minimum.
2 Bypass and grid lines are as short as possible.
3 Longer lines, such as ground and heater circuits, are placed near or around the edge of the board.
4 The design is neat and compact.

If the above conditions are not satisfied, the drawing may have to be revised or a new sketch might have to be made in which the components would be relocated or reoriented. Figure 7·4 shows such a sketch of the circuit shown in Fig. 7·3. The transistors have been sketched as circles, and the other components as rectangles or squares. The transistors have been regrouped so that the biasing resistors can be located closely, and the only long lines are the ground, +15-volt, and −15-volt paths. Resistors $R6$ and $R7$ and diodes $CR1$ and $CR2$ (with their polarities carefully noted) are con-

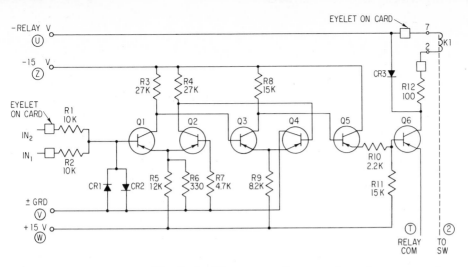

Fig. 7·3 Elementary diagram of a relay-comparator amplifier. (Electronics Associates, Inc.)

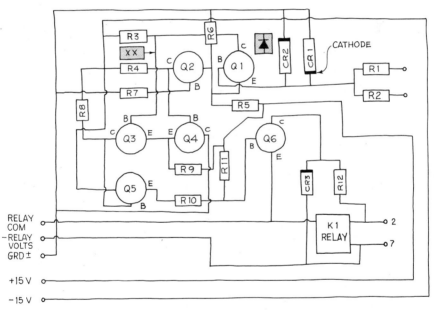

Fig. 7·4 Preliminary sketch for placement and printed wiring of components or circuit shown in Fig. 7·3.

nected to the common ground. At first inspection there appear to be many crossovers, as with line *xx*, for example. However, these crossovers can be eliminated by passing the line directly under the components (let line *xx* go under *R*4 and

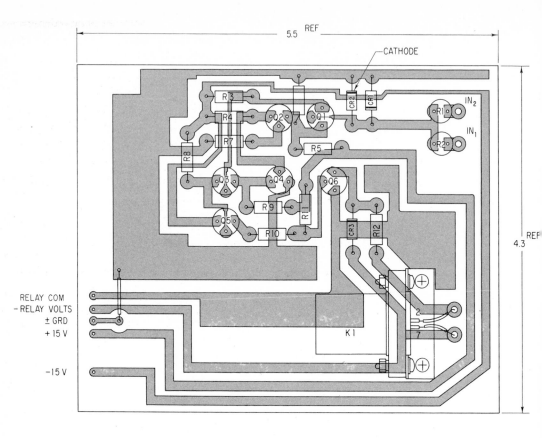

Fig. 7·5 Drawing for assembly of components, made from master layout. (From a drawing by Electronics Associates, Inc.)

R7.) If this is to be a single-coated board, there will be no circuitry at these locations and the component will be on the other side. Figure 7·5 illustrates this situation quite well.

Once a satisfactory sketch, such as that shown in Fig. 7·4, is obtained, it should be checked for correctness. A breadboard model of the circuit can even be constructed and tested. The next step is to make a *preliminary layout* for the master drawings. This is often made on a standard grid, such as appears in Fig. 7·2, and to scale. Scales of 2:1 and 4:1 are convenient to work with. (The grid spacing must be multiplied

5.500

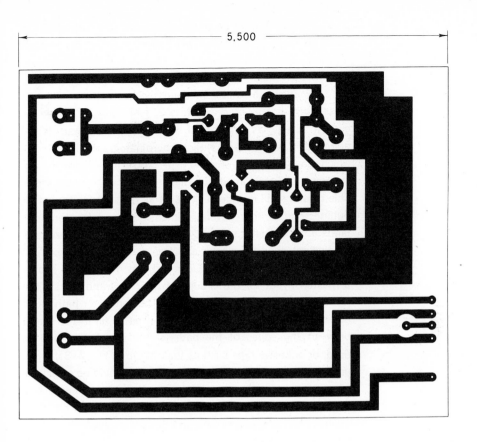

Fig. 7·6 Master layout of printed circuit board for relay-comparator amplifier. (Electronics Associates, Inc.)

by the scale factor.) Many components that conform to standard grid spacing and that can be mounted directly on such a board are available and should be utilized for this type of design. They are usually aligned horizontally or vertically but may be located in other positions if the requirements so dictate. The drawing, at this point, will look very much like that of Fig. 4·18. After the components and hole centers are located, the circuit-path pattern can be drawn in, either on the same sheet—usually on the back—or on another overlay. Physical features such as fasteners and jumpers, as well as terminal pads, should be shown. This drawing should also be carefully checked because it will be the basis for the master layout.

Now the *master layout* can be drawn. It should be drawn on a dimensionally stable medium such as polyester film or glass cloth. It may be drawn to a 2:1, 4:1, or 8:1 scale or even to a scale as large as 20:1.* If extreme accuracy is required, it can be made with a coordinatograph (Fig. 7·7), which has a plotting accuracy of 0.001 in. Otherwise, an accurate undergrid made to the proper scale can be used. This drawing will include such items as register points, board outline, one or more overall dimensions, and the terminal lands and conductor pattern. The terminal pads are generally drawn first, then the elbows and tees, and finally the conductor paths between elbows, pads, and tees. Figure 7·6 shows a master layout for the circuits of Figs. 7·3 and 7·4. (This layout is the underside of Fig. 7·4.) The conductor path must be solid black. A special ink can be used to fill in the space between the edges of the paths if the coordinatograph is used, or adhesive tape such as that shown in Fig. 7·8 can be used for less accurate requirements. Care should be exercised when using tape because of its tendency to "creep," but with proper use, it and the other appliqués shown in Fig. 7·8 will save much drafting time and provide a nice-looking pattern. Another accurate system for drawing conductor paths uses a scribe that has two points, accurately spaced, that cut sharp, precise lines on a peel-coat polyester film. The area between the two lines represents the path and remains after the drawing is developed and washed, following which the remainder of the film is peeled away. Component designations and other items of information may be placed on the master or may be placed on a separate marking drawing made from the master. Lettering and register marks should be at least 0.015 in. thick after reduction, and letters should be tall enough to read after reduction to final size.

From the master layout, other drawings are often made. These would include the *drilling drawing*, which might be similar to that of Fig. 4·17, a *component assembly* drawing such as that shown in Fig. 7·5, *a marking drawing*, not

*Other processes such as photography place a limit on the overall size of such a drawing, however. This is about 20 × 30 in.

Fig. 7·7 A coordinatograph which makes possible precision drafting for printed circuits and other miniaturized parts. (Keuffel & Esser Co.)

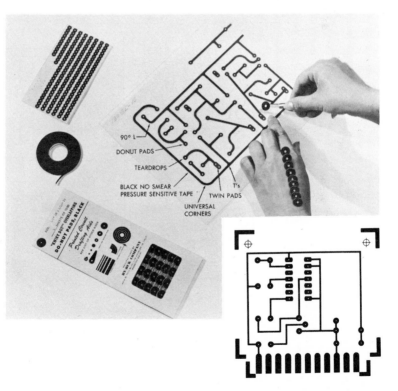

Fig. 7·8 Adhesive aids for printed circuits layouts. (BYBUK Company and Bishop Graphics, Inc.)

shown, and possibly mechanical-assembly drawings. It is also possible to buy printed resistors, capacitors, wiping switches, and a few other printed components for use in printed circuits.

The following items are from a checklist suggested by a company that specializes in making printed-circuit drawings:

1 Is the drawing made on polyester-based material? Paper and illustration board must be avoided.
2 Have drilling or spotting guides been included? An adequate guide is provided by a white dot in the center of the black land pattern with a final size of 0.02 in.
3 Have 90° guidelines been included? These are essential to accurate positioning in the step-and-repeat photocomposing machine.
4 Has there been specified just one critical dimension to which the master drawing is to be reduced? This is a photographic necessity.
5 Is the land pattern around each hole at least 0.31 in. in

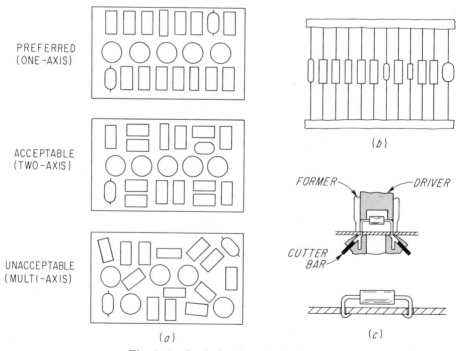

PREFERRED
(ONE-AXIS)

ACCEPTABLE
(TWO-AXIS)

UNACCEPTABLE
(MULTI-AXIS)

(a)

(b)

FORMER   DRIVER

CUTTER
BAR

(c)

Fig. 7·9 Optimization of PC design. (a) Preferred and unacceptable design for NC assembly. (b) Lead taped components. (c) Schematic drawing of NC insertion machine action.

width? The diameter of the copper pad must be at least ⅛ in. larger than the hole size.

6 Is the master drawn in ink? Are the edges of the lines clean and sharp?

7 Have sharp corners on conductor paths been avoided?

8 Has spacing between conductors or lands on circuits to be plated been increased at least 0.01 in. over the required minimum?

Some of these items do not apply to the printed circuit used here, but they do represent good practice, in general.

In order to lower prices (or to keep them from rising because of increasing labor costs) automation of the printed-circuit-board assembly has been taking place. The numerical control (NC) insertion machine automatically puts components on a board and clinches them in position as shown in Fig. $7 \cdot 9c$. This method of assembly requires that the pattern be limited to X and Y axes as shown in Fig. $7 \cdot 9a$. The components have been previously put in lead tapes in the order in which they are to be inserted on the board. The tape moves through the machine much as an ammunition belt is fed through a machine gun. Components are then placed in the board one after the other in a line. This is why the one-axis design is preferred. The two-axis method will work but requires more passes and is slower. However, it must be remembered that the circuit pattern must be designed to accomodate these requirements, and this might be extremely difficult—if not impossible—for some circuits.

7·4 Integrated circuits

Integrated circuits may be classified as follows: (1) thin film, and (2) monolithic silicon. Classification can be further complicated by mentioning thick-film and hybrid circuits, but, in general, there are two types of ICs, depending upon how and of what materials they are made.

Integrated circuits make it possible to form many transistors, diodes, resistors, and other components within a single structure the size of which might be less than a typical transistor can or slightly larger. Figure $7 \cdot 10a$ shows two thin-film integrated circuits that have the components shown in Fig. $7 \cdot 10b$. Figure $7 \cdot 11$ shows a module containing ⅛ in. square chips, each of which has 174 complete memory circuits.

Fig. 7·10 Thin-film circuits. (a) Two circuits mounted on a TO-5 header. (b) Physical electrical schematic of logic gate shown in (a). (Copyright 1965, The American Telephone and Telegraph Company. Used by permission.)

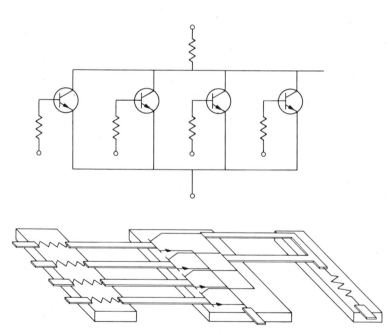

Fig. 7·11 Memory module for the IBM System/370 computers. Four small chips are mounted on the double-layer module. Each monolithic chip has 1,400-plus elements. (International Business Machines Corp.)

Fig. 7·12 Integrated-circuit memory unit that is electrically programmable. It is an MOS (metal oxide) circuit which fits in a dual in-line package. (INTEL Corp.)

Figure 7·12 shows a greatly enlarged photograph of a memory unit that has 256 words and 2,048 bits of memory within a rectangular area about ½ × 1⅛ in. These semiconductor IC (integrated circuit) memories are supplanting core and drum memories for several reasons, one being size (about one-half the volume of comparable core memory units) and another cost (lower cost per "bit," or binary digit, than the core units).

(a)

(b)

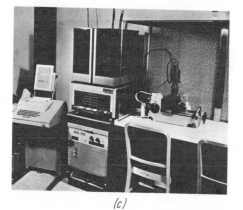

(c)

Fig. 7·13   Equipment used in production of integrated circuits. (a) An automatic drafting machine. (Gerber Scientific Instrument Co.) (b) Trimming thin-film resistors to desired size with an abrasive trimming machine. (Pennwalt S.S. White Industrial Products). (c) Computer-controlled step-and-repeat camera. (Copyright 1968, Bell Telephone Laboratories, Inc. Used by permission.)

Volume production of these microminiature circuits requires expensive equipment. Some of this equipment is shown in Fig. 7·13. The automatic drafting machine can be used to accomplish two objectives: (1) to render the most accurate drawing possible, and (2) with the computer to design some of the circuits (for reasons that will be explained later). The abrasive trimmer is used to trim resistors and capacitors of thin-film circuits to the exact size (to get the exact resistance, for example). Step-and-repeat photographic equipment is used to reduce the original drawings (often referred to as artwork and masks) by factors of 200 or less. Reduction has to be done by steps and with fine lenses in order that distortion be negligible.

7·5 Thin-film circuits

Thin-film circuits are those in which electrical components are deposited in thicknesses up to a few microns (1 μ = 0.0001 cm) on a supporting substrate of glassy or ceramic dielectric. Typical thin-film materials for resistors and capacitors are:

resistors	*capacitors*
Tantulum	Tantulum oxide
Nickel chrome	Titanium oxide
Titanium	Thin ceramic layers
Cadmium sulfide	Silicon dioxide

Two preferred methods for depositing the film are vacuum evaporation and cathode sputtering. Patterns can be generated by means of photoetching techniques. The resistors may be brought to their final value by a process known as *trim anodizing*. Resistors are also covered by oxide for protection. Contacts, or contact areas, may be made of nickel chrome covered by gold.

Figure 7·14 shows a photograph of a thin-film circuit that is 2 in. long. Figure 7·15a is a sketch of that circuit in which some of the features are explained. Note that the contact areas are composed of nickel chrome (NiCr) plated with gold, all over tantulum. The four resistors shown at $R5$ (Fig. 7·15b) produce about 65 ohms resistance, and because they are in

Fig. 7·14   Photograph of a thin-film circuit. (Western Electric Co.)

parallel, their individual resistances must be divided by 4.

Figure 7·15*b* shows the completed drawing for this circuit. The five main items are:

1   Elementary diagram
2   Scale drawing of thin-film circuit (5:1)
3   Calculation of the areas of the resistors
4   Table for resistor sizes and capacities
5   Table for capacitor sizes and capacities

The thin-film devices currently in use contain mostly *passive* devices, such as capacitors and resistors. It is also possible to make some diodes and transistors (active devices) in this way. Inductors, however, are impossible to make because of the incompatibility of the geometry.

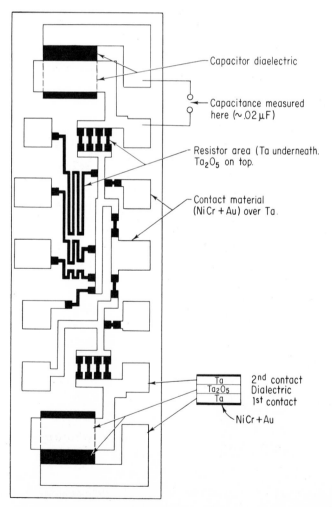

Fig. 7·15   Thin-film structure. (a) Sketch of a thin-film-circuit structure.

Fig. 7·15 (b) Scale drawing and calculations for the manufacture of a thin-film circuit. (Western Electric Co.)

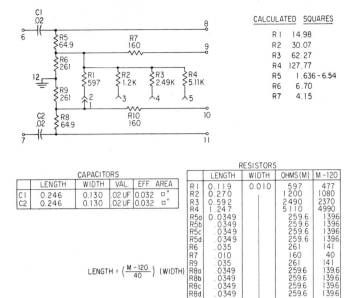

CALCULATED SQUARES

R1	14.98
R2	30.07
R3	62.27
R4	127.77
R5	1.636 - 6.54
R6	6.70
R7	4.15

CAPACITORS	LENGTH	WIDTH	VAL.	EFF. AREA
C1	0.246	0.130	.02 UF	0.032 □"
C2	0.246	0.130	.02 UF	0.032 □"

$$\text{LENGTH} = \left(\frac{M-120}{40}\right)(\text{WIDTH})$$

RESISTORS

	LENGTH	WIDTH	OHMS (M)	M-120
R1	0.119	0.010	597	477
R2	0.270		1200	1080
R3	0.592		2490	2370
R4	1.247		5110	4990
R5a	0.0349		259.6	1396
R5b	.0349		259.6	1396
R5c	.0349		259.6	1396
R5d	.0349		259.6	1396
R6	.035		261	141
R7	.010		160	40
R9	.035		261	141
R8a	.0349		259.6	139.6
R8b	.0349		259.6	139.6
R8c	.0349		259.6	139.6
R8d	.0349		259.6	139.6
R10	.010		160	40

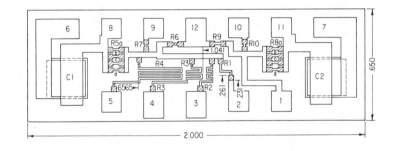

It has been necessary to turn to the computer for help in getting smaller components, thinner line widths, and more accuracy and reliability. The resistors in the circuit shown in Fig. 7·16 were designed and drawn by a computer. The computerized mask-making system for this type circuit has eliminated much manual drafting and cut production time significantly. Figure 7·17 shows the artwork and software that were connected with another thin-film circuit.

7·6 Integrated (or monolithic silicon) semiconductor circuits

Integrated-semiconductor-circuit techniques enable engineers to develop the equivalent of an entire assembly of components on a small silicon wafer. In this process, the wafer is highly polished, then masked by a layer of oxide or other substance.

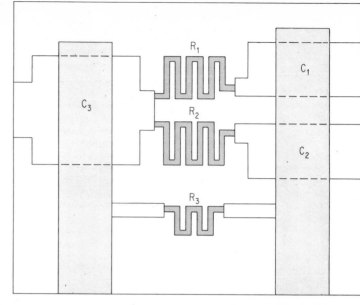

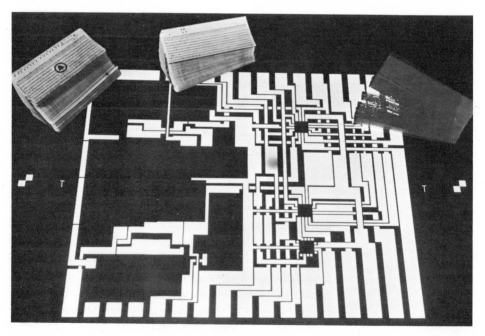

*Fig. 7 · 17 Artwork for a thin-film-resistor network. The
masks have been generated by computer-controlled equipment. This circuit had 1,400 coordinates. (Copyright 1969, Bell
Telephone Laboratories, Inc. Used by permission.)*

Transistors, diodes, and other components are formed on the wafer's surface by opening tiny windows and driving selected impurities or gases into the openings. A pattern of conducting material is then plated over the entire circuit, connecting its elements. As many as several hundred circuits can be formed at the same time, then cut apart.

Figure 7·18 shows the steps in the construction of a planar transistor. Donor impurities, such as phosphorus and arsenic, are put into the silicon by means of the diffusion process to form the N region. Acceptor impurities, such as boron and gallium, are impregnated in the silicon by the same process to form the P region. Another concept is to deposit a thin film of a semiconductor material on the substrate by a gas or vapor process. This process is called *epitaxial growth*.

Accurate surface geometry enables the manufacturer to obtain fine-scale patterns, both for the diffusion of impurities and for the deposition of the intraconnection metallization pattern. The formation of an integrated circuit might require from five to eight masking steps. Silicon dioxide (SiO_2) makes a very effective masking agent for diffusants. Diffused paths will be formed only where the SiO_2 has been selectively re-

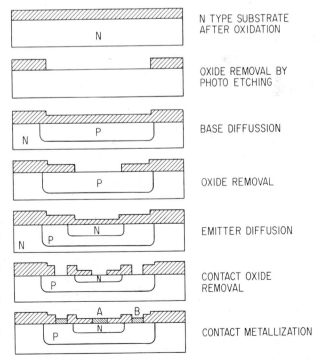

Fig. 7·18 Schematic cross sections of the steps necessary to make a single planar PNP transistor on a silicon wafer. (From Edward Keonjian, "Microelectronics," McGraw-Hill Book Company, New York, 1963. Used by permission.)

N TYPE SUBSTRATE AFTER OXIDATION

OXIDE REMOVAL BY PHOTO ETCHING

BASE DIFFUSSION

OXIDE REMOVAL

EMITTER DIFFUSION

CONTACT OXIDE REMOVAL

CONTACT METALLIZATION

moved. The pattern is usually repeated by step-and-repeat photography until the final pattern contains many replicas of the original drawing. The mask containing these many circuit patterns is placed on the wafer by photolithography.

Figure 7·19 shows a schematic cross section of a wafer in which various elements have been formed. Figure 7·20 shows how some typical patterns might appear during three diffusion steps, as one looks at right angles to the surface of the wafer. There are eight capacitors, seven diodes, nine resistors, and seven transistors. With these, six different networks have been formed using different patterns of intraconnection. Two of the networks which have been formed *from the same pattern* are shown in Fig. 7·21.

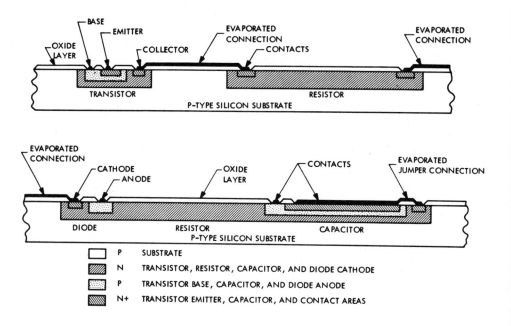

Fig. 7·19 *Cross section of semiconductor network wafer showing diffusions and technique used to connect component areas to form desired circuit.* (*Texas Instruments Incorporated.*)

7·7 Drawings for monolithic integrated circuits

Figures 7·22 to 7·26 show the drawings which are necessary to make and connect a typical integrated circuit. Figure 7·22 shows a scale drawing that outlines the entire package, the piece of silicon itself, and the patterns of the components themselves. Notice the grid system that is used to facilitate accuracy. Spacing between components is actually as close as 0.0005 in. at times; therefore, alignment errors cannot

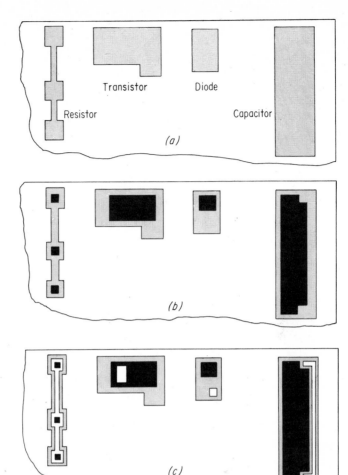

Transistor Diode

Resistor Capacitor

(a)

(b)

(c)

Fig. 7·20 Surface of semiconductor network as it would appear during three diffusion steps, (a), (b), and (c). Scale about 40:1.

exceed more than a few ten-thousandths of an inch. This diagram was drawn to a scale of 150 times the actual size, although it has been reduced several times to fit on the pages of this book. Figure 7·23 shows the second drawing, which indicates the first diffusion pattern and which is called a *bar layout.* Figure 7·24 shows another in the series of diagrams. This drawing shows the evaporated aluminum connections, but it looks very similar to the third diffusion pattern. Figure 7·25, called a *bonding diagram,* shows the same diagram as Fig. 7·24, but the ends of the Kovar leads are shown. With this diagram, the design engineer makes a drawing to indicate how he wishes the components to be interconnected.

Fig. 7·21 Elementary diagrams of two circuits formed from the same diffusion pattern. (a) Input NOR/NAND *gate. (b)* EXCLUSIVELY OR *gate. (Texas Instruments Incorporated.)*

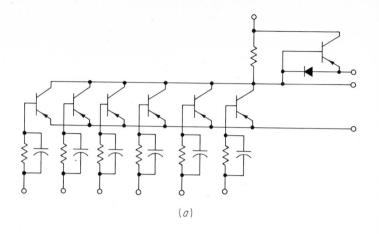

(*a*)

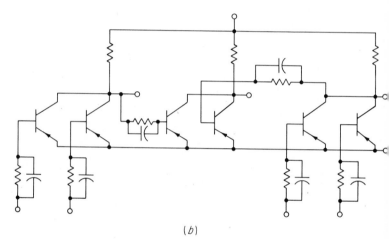

(*b*)

Then a finished intraconnection diagram is made to prepare a mask. In production, aluminum is evaporated over the entire surface of the slice. The aluminum intraconnection patterns are formed by etching away the aluminum using a photolithographed copy of the intraconnection diagram. An intraconnection diagram for a slightly different type of integrated circuit than we have been using so far is shown in Fig. 7·26. The circular disks represent the exterior leads in this diagram. Note that the final mask provides for manufacture of 157 identical networks on an area of silicon substrate about 1 in. square. The photograph of a mask for an intraconnection diagram for another IC is shown in Fig. 7·27.

Fig. 7·22 The first of a series of drawings for the manufacture of an integrated circuit or network. The outer rectangle represents an outline of approximately $\frac{1}{8} \times \frac{1}{4}$ in. (Texas Instruments Incorporated.)

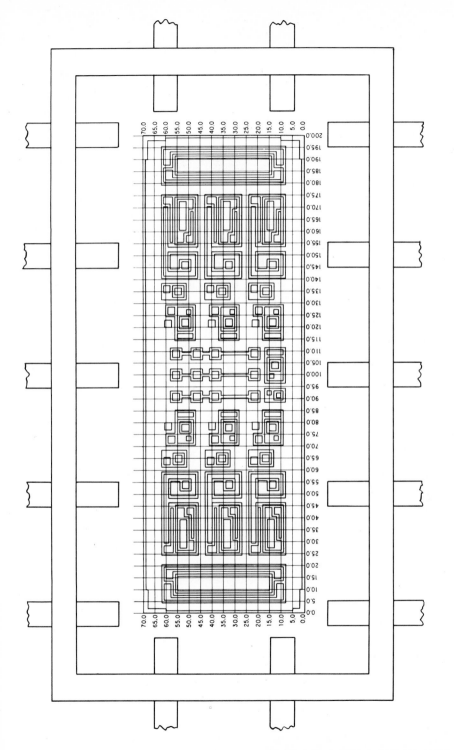

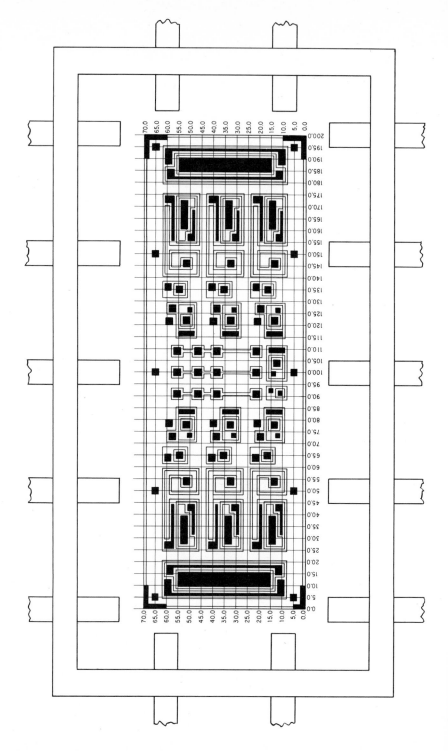

Fig. 7·23 The second drawing of a series for an inte-grated network. There are eight capacitors, seven diodes, nine resistors and seven transistors on this drawing. (Texas Instruments Incorporated.)

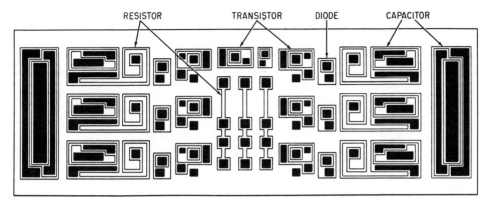

Fig. 7·24 Another in the series of drawings for an inte-grated circuit.

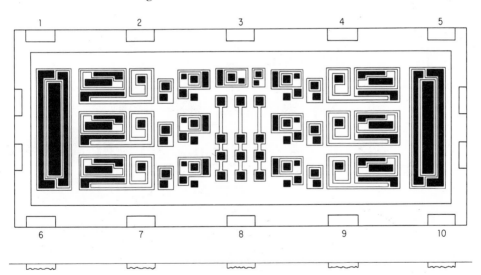

Fig. 7·25 The next drawing in the series. Leads have been added and numbered. (Texas Instruments Incorpo-rated.)

Figure 7·28 is a debugging drawing consisting of outline draw-ings of nine masks all made by a computer-prepared tape.

Finally the network must be separated from other like net-works by a suitable, accurate cutting process and then assem-bled and packaged. Figure 7·29a shows the construction and

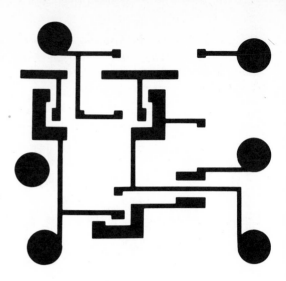

Fig. 7·26 An intraconnection diagram and final working mask of the same pattern. (From Edward Keonjian, "Microelectronics," McGraw-Hill Book Company, New York, 1963. Used by permission.)

← 157 patterns (identical)

Fig. 7·27 Mask for an interconnection diagram for a monolithic integrated circuit. (Copyright 1969, Bell Telephone Laboratories, Inc. Used by permission.)

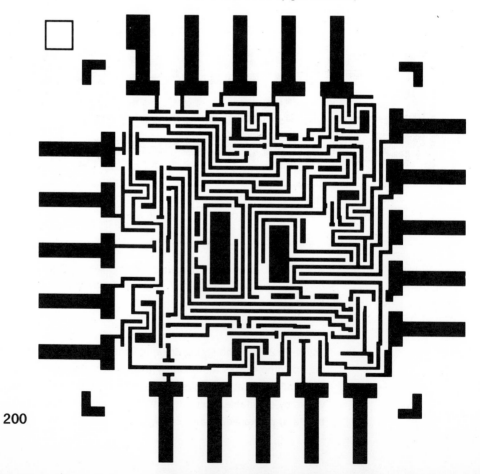

an outline of a flat package with 14 leads. The bottom of the package is metallic, acting as a heat sink if desired. The cap (lid) is welded to the package, and the leads and circuit are insulated from it electrically. Drawings of a transistor "can" which is used for the packaging of integrated circuits appear in Fig. 7·29b. As in the case of flat packages, circular packages may come with different numbers of pin connections. The third type of package, shown in Fig. 7·29, lends itself well to modern packaging design. (See Figs. 7·2 and 7·9.) A large number of off-the-shelf semiconductor integrated circuits, as well as numerous "custom" circuits, are manufactured.

7·8 Hybrid circuits

Thin-film and monolithic silicon integrated circuits can be combined. There are several reasons for making these *hybrid* circuits. It is not usually feasible to make transistors by thin-film techniques. Therefore, if it is desired to have thin-film patterns or networks in which transistors or diodes are required, these *active* elements will have to be added—possibly in discrete form—to the film network. (See Fig. 7·31.)

Another reason for combining different types of circuits is that the range of manufacture, tolerance, and temperature coefficients of certain devices is more limited in one process than in another. Tables 7·4 and 7·5 show some of these limitations. It will be quickly noted that the manufacturing

Table 7·4   Characteristics of resistors

	major range of values		
	resistance (ohms)	*manufacturing tolerance*	*temperature coefficient of resistance (ppm/°C)*
Integrated semiconductor circuits (diffused silicon)	100 ↓ 50K	±5% ↓ ±20%	50 ↓ 5000
Thin films	10 ↓ 10MEG	±0.01% ↓ ±10%	−25 to +150 ↓ −300 to +300

Table 7·5 Characteristics of capacitors

	breakdown voltage (volts)	capacitance (pfd/mil²)	mfg. tolerance
Integrated semi- conductor circuits (diffused silicon)	10	0.45	±15%
	30	0.30	
	70	0.04	±50%

	material	working voltage	capacitance (pfd/mil²)	mfg. tolerance
Thin films	SiO	6	3.2	±10%
	SiO	~80	0.045	±10%
	SiO_2	50–100	0.026	±10%
	Ta_2O_5	~100	0.65	±3%
	TiO_2	50	1.65	±2%

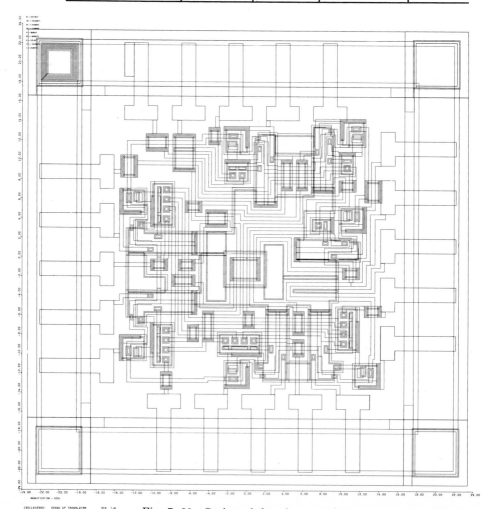

Fig. 7·28 Series of drawings made by a computer. These "debugging plots" are composite views of all nine masks which have been superimposed. (Copyright 1969, Bell Telephone Laboratories, Inc. Used by permission.)

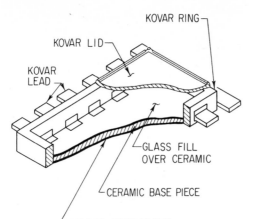

KOVAR RING

KOVAR LID

KOVAR LEAD

GLASS FILL OVER CERAMIC

CERAMIC BASE PIECE

WELDED KOVAR PLATE

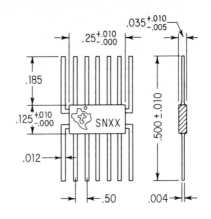

$.25^{+.010}_{-.000}$

$.035^{+.010}_{-.005}$

.185

$.125^{+.010}_{-.000}$

SNXX

$.500 \pm .010$

.012

.50

.004

(a)

10 PIN TO-5

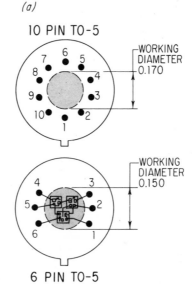

WORKING DIAMETER 0.170

WORKING DIAMETER 0.150

6 PIN TO-5

(c)

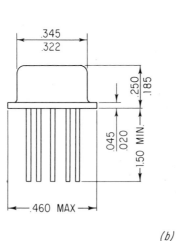

.345
.322

250
.185

.045
.020

1.50 MIN.

.460 MAX

(b)

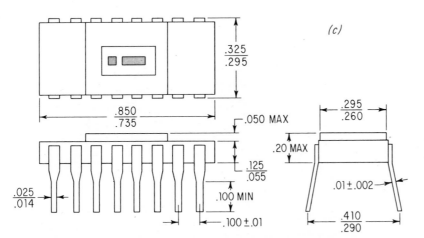

.325
.295

.850
.735

.050 MAX

.125
.055

.025
.014

.100 MIN

.100 ± .01

.295
.260

.20 MAX

.01 ± .002

.410
.290

Fig. 7· 29   Packages for monolithic integrated circuits. (a) Flatpack. (Texas Instruments Incorporated.) (b) TO-5 packages. Several IC chips are shown attached in the lower figure. (c) Dual in-line (DIP) package. (INTEL Corp.)

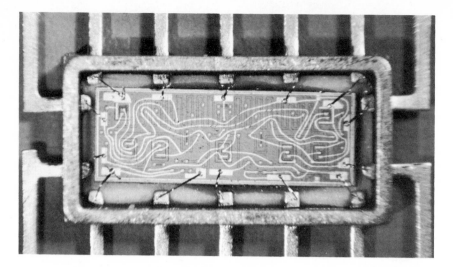

Fig. 7·30 Photograph of an integrated semiconductor circuit with connections made prior to encapsulation. (Texas Instruments Incorporated.)

tolerances of both resistors and capacitors are lower (closer) in thin-film techniques than in the diffusion process performed on silicon wafers. Thus, it is possible, and sometimes desirable, to form another type of hybrid circuit with silicon chips (or *blocks*) and thin-film techniques.

In this type of circuit (see Fig. 7·32), active components are formed within a silicon block and the passive-component pattern is deposited by thin-film techniques on top of a passi-

Fig. 7·31 A frequency divider network in which microminiature "active" devices in discrete form have been added to a thin-film pattern. (Motorola Semiconductor Products Division.)

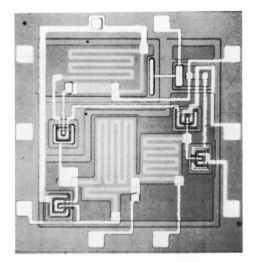

Fig. 7·32 Another type of hybrid circuit. Thin-film patterns of passive elements have been created on top of the insulated surface of a semiconductor active network. The diagram at the bottom is an exploded-view drawing of another typical circuit of this type. (Motorola Semiconductor Products Division.)

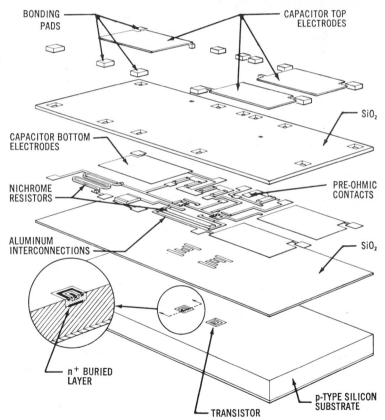

BONDING PADS

CAPACITOR TOP ELECTRODES

SiO₂

CAPACITOR BOTTOM ELECTRODES

NICHROME RESISTORS

PRE-OHMIC CONTACTS

ALUMINUM INTERCONNECTIONS

SiO₂

n⁺ BURIED LAYER

p-TYPE SILICON SUBSTRATE

TRANSISTOR

vating layer (SiO₂) which covers the "active" circuit. This hybrid circuit (called a "compatible" circuit by its manufacturer) utilizes the advantages of both monolithic Si and thin-film processes. Because it requires more manufacturing steps, it is more expensive than the monolithic semiconductor or thin-film circuits. Occasionally, the thin-film network is too large to fit upon the silicon chip, and special interconnecting and manufacturing processes have to be used. It is also possible to use printed circuitry with integrated and thin-film circuits. In many cases, this is more a matter of incorporating integrated circuits into an overall pattern or system than hybridizing as we have just discussed it.

Fig. 7·33 Another hybrid circuit. Contacts and connections are nearly white in color. Thin-film elements (medium to dark grey) can be seen superimposed upon the active elements. (Fairchild Semiconductor Division.)

SUMMARY The development of transistors and etched circuits brought great benefits in the reduction of size and improvement of reliability of electronics circuits and their packaging. Random geometry, uniform geometry, and hybridization are natural allies of printed circuitry.

Thin-film circuits using, for the most part, passive devices carry the printing concept to the point where the devices themselves are printed. Monolithic semiconductor (solid-state) circuits utilize the printing concept also, but in addition require such processes as diffusion and evaporation. Several production steps are performed on specially prepared germanium or silicon wafers.

Because of the closeness of elements in miniature or microminiature circuits, tolerances are very small. Drawings, then, must be extremely accurate. This is achieved by making the drawings several, or many, times larger than the finished product and using special tools and techniques. A thorough understanding of production methods is required of anyone engaged in the making of such drawings. In many cases, a good understanding of circuits is required.

Integrated circuits can also be classified as MSI and LSI. MSI (medium-scale integration) chips have less than 100 transistor functions. LSI (large-scale integration) chips include hundreds of transistor functions. LSI chips have made the "pocket" calculator possible.

QUESTIONS

7·1   What are the advantages of thin-film circuits over integrated silicon circuits?

7·2   Why are one- or two-axis designs preferred for circuit boards?

7·3   What drawings would be required for the manufacture of a thin-film circuit? What else would probably be included in the drawings?

7·4   What are four specific problems brought on by the problem of great numbers of components?

7·5   What are the main approaches in use in miniaturization today?

7·6   List four scales (other than full-size) to which drawings for miniature or microminiature circuits have been drawn.

7·7   Why is the automated drafting machine sometimes used for drawings of integrated circuits?

7·8   What drawings may be required for the production of a rather complicated printed circuit?

7·9   What are some typical widths for printed conductor paths?

7·10   What is a good minimum distance between an outside conductor and the outer edge of a printed-circuit board?

7·11   What is the first step in making drawings for printed circuits?

7·12   Why is a meandering line thin-film resistor not the same in area as its straight-line equivalent?

7·13   What is your concept of a matrix?

7·14   What is your concept of the following terms, as they apply to miniature circuits or microcircuits?
a. Random geometry
b. Surface geometry
c. Uniform geometry

7·15   What material or materials are generally used as substrate(s) for integrated (monolithic) circuits?

7·16 What is an effective masking agent for the above? Name an acceptor impurity. A donor impurity.

7·17 How is it possible to get a series of six or eight circuits from one integrated semiconductor pattern having capacitors, diodes, resistors, and transistors?

7·18 What are four drawings, of a series, required for production of a completed, integrated (semiconductor) circuit package? What scales would be appropriate for these drawings?

7·19 Where does the inductor come into the thin-film and integrated-semiconductor-circuit picture? Would this have an effect on circuit design?

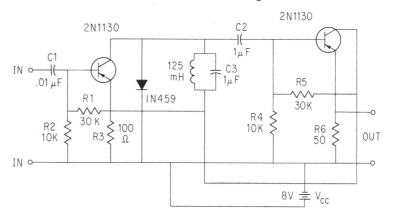

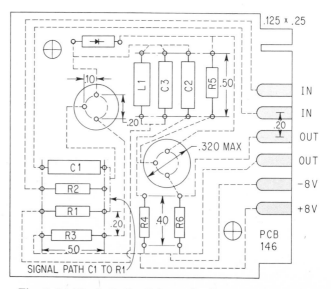

Fig. 7·34 (Prob. 7·1) Schematic diagram and preliminary drawing of circuit board for a current-sweep generator.

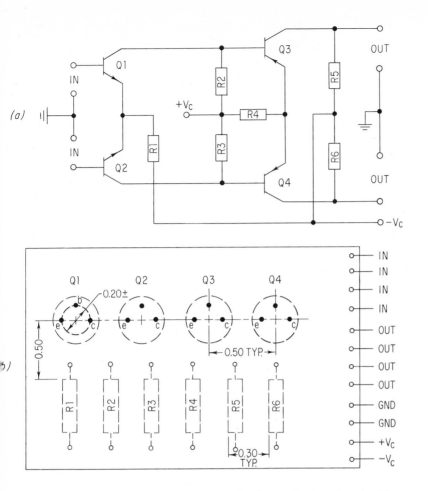

(a)

b)

Fig. 7·35 (Prob. 7·2) A printed-wiring problem for a differential-amplifier circuit.

7·20 What is your concept of hybridization in so far as this chapter is concerned?

7·21 What trends do you identify in electrical drawing for small, miniature circuits and devices in the next decade?

7·22 What are some of the uses of a master layout in printed circuitry?

7·23 Show, by means of sketches, the following: eyelet, jumper, pad, registration mark, board outline, connector(s) used in printed-circuit plug-in boards.

7·24 Show, by means of sketches, the following: thin-film resistor (two shapes), thin-film capacitor, diode (integrated circuit), transistor (integrated circuit), and DIP.

7·25 What is meant by passive elements? By active elements?

PROBLEMS 7·1 Figure 7·34 shows a preliminary layout of a circuit board that is to contain the sweep-meter circuit shown. Check the layout against the schematic diagram for possible errors or improvement in layout. Make a PCB layout (outside dimensions are 2.20 × 2.40 in.) to a 4:1 or 5:1 scale. Use the 0.10 grid system for locating components, whose sizes are: resistors, .25 × .09, except $R3$ which is .375 × .09; capacitors, .422 × .135; $L1$, .400 × .15; and diode, .275 × .105 max. Use .062 conductor paths, with .031 minimum spacing. Use 8½ × 11 paper.

7·2 Figure 7·35 includes an elementary diagram a and a suggested arrangement of parts for a differential amplifier. Make a scale drawing (4:1 or 5:1) of the final master drawing for the pattern. A maximum of four jumpers will be permitted. The lower figure shows the wiring side of one arrangement of components. Transistor diameters are .360 ±.010, and resistor dimensions are .093 dia. × .375. Try to get an acceptable pattern on a board that is no larger than 1.50 × 2.00 in. and is .062 thick. Freehand preliminary sketches are suggested. If you cannot get an acceptable pattern with the suggested arrangement of components, make your own arrangement. Hole diameters are .032. Terminal pads are .125 in.

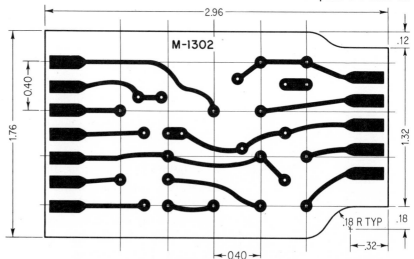

Fig. 7·36 (Prob. 7·3) Printed-circuit pattern.

across. Use .062-in. conductor paths with .031 minimum spacing. Show registration marks, pin "A" (top pin), and label the board PCB 46. Use 8½ × 11 paper. Show one critical dimension.

7·3   A printed-circuit pattern is shown in Fig. 7·36. Make a master layout drawing of the same pattern at 4:1 or 5:1 on a 0.10-in. grid. (Some holes may have to be relocated to fit on the intersecting grid lines. Use .062-in. conductor patterns with .031 minimum spacing, including distance to edge of board. Holes are .032″, donuts 0.125″. Show board number, registration marks, and one principal dimension. If tape and

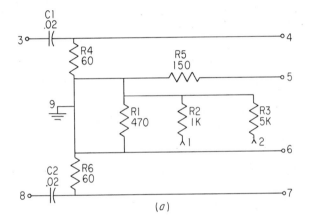

(a)

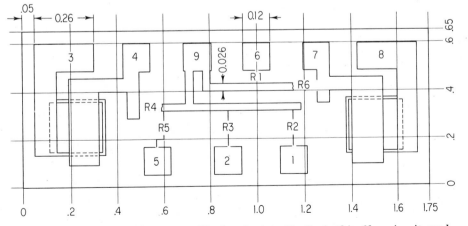

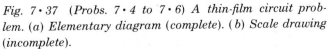

Fig. 7·37 (Probs. 7·4 to 7·6) A thin-film circuit problem. (a) Elementary diagram (complete). (b) Scale drawing (incomplete).

adhesives are to be used, circuit paths may be altered (straightened) for easier use of tape. Use 8½ × 11 paper.

7·4　Figure 7·37 shows the elementary diagram of a thin-film circuit and a scale drawing, complete except for the resistors. Given resistor material of 0.01-in. width, compute the length for the resistors using the equation $L = [(M - 120)/40]W$. Then make a table for the resistors and a calculated-squares table as they appear on Fig. 7·15.

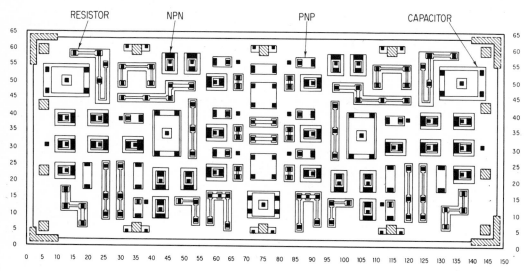

Fig. 7·38　(Probs. 7·7, 7·8) A bar diagram of an integrated semiconductor circuit. (Texas Instruments Incorporated.)

7·5　As a continuation of Prob. 7·4, complete the scale drawing of the thin-film circuit shown in Fig. 7·37. Use the resistor designs that provide the lengths (or resistances) you calculated in the previous problem. Typical dimensions and the 0.2 grid system should enable you to draw a reasonably accurate scale drawing at a scale of 5:1. Use 8½ × 11 paper.

7·6　Do the calculations required in Prob. 7·4, and show the tables along with the given elementary diagram of the thin-film circuit and the completed scale drawing (Prob. 7·5) on one sheet in a format similar to that shown in Fig. 7·15. Use 11 × 17 or 12 × 18 paper.

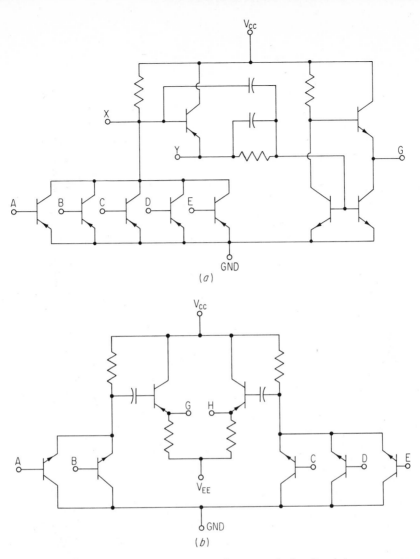

Fig. 7·39 (Prob. 7·8) Integrated circuits. (a) AND *or* NOR *gate. (b)* AND *or* OR *gate.*

7·7 Figure 7·38 shows a bar diagram of an integrated circuit. Count the PNP transistors, NPN transistors, capacitors, and resistors in this circuit and make a table of your results. Do not count configurations which are not specifically like the PNP and NPN transistors located. Resistors have several shapes and vary from 0.3K to 8K for a total of 70K. Capacitors are either 25PF or 15PF. Hint: There are more than 30 transistors. Use 8½ × 11 paper.

7·8 Figure 7·39a and b are elementary diagrams of two different circuit configurations which can be made

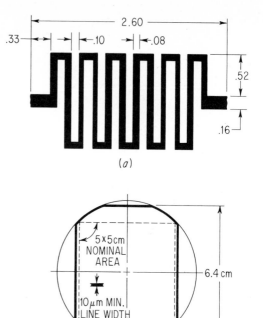

Fig. 7·40 (Prob. 7·9) Details of thin-film drawings. (a) Meander resistor with defects. (b) Standard 5-cm field.

(a)

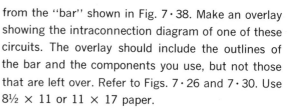

(b)

from the "bar" shown in Fig. 7·38. Make an overlay showing the intraconnection diagram of one of these circuits. The overlay should include the outlines of the bar and the components you use, but not those that are left over. Refer to Figs. 7·26 and 7·30. Use 8½ × 11 or 11 × 17 paper.

7·9 One of the problems of manufacturing microcircuits is that of accuracy. Figure 7·40a shows a photographically enlarged resistor which has lines of differing widths because of inaccuracies in the original drawing and manufacturing process. This resistor is to have a resistance of 7540 ohms and a line width of .08 as shown. Compute the length, using the formula $L = (R - 140)/40 \ W$, where the 140 takes care of "cornering" losses. Make a drawing to an enlarged size of the resistor using the dimensions shown and the length you have computed. Use 8½ × 11 paper.

7·10 Figure 7·41 shows three assorted views of a case for

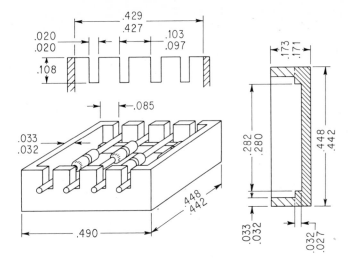

Fig. 7·41 (Prob. 7·10) Plastic case for matched-quad diode assembly: pictorial view, slot details, and a section.

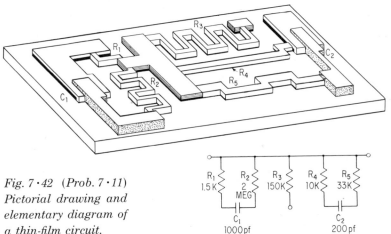

Fig. 7·42 (Prob. 7·11) Pictorial drawing and elementary diagram of a thin-film circuit.

R_1	R_2	R_3	R_4	R_5
1.5K	2 MEG	150K	10K	33K

C_1 1000 pf

C_2 200 pf

a matched-quad diode assembly. Make a working drawing of the molded plastic case for the assembly. A drawing to a size of 4:1 or larger is recommended. Give this a drawing number 10047-4, max voltage 180V, and max current 7A.

7·11   Figure 7·42 shows a pictorial sketch and elementary diagram of a thin-film circuit. Using a width of 0.01

in. or wider, compute the length of the resistors and make a scale drawing of the final circuit. Refer to Figs. 7·15 and 7·37. Drawing may include the scale drawing, schematic diagram, and length calculations.

Chapter 8

Industrial controls and automation

There is enough material on industrial controls to fill a book—in fact such books have been written—and we shall be able to show only a few isolated examples in this chapter. Because many types of industrial controls are not electronic, some of the circuits that are included are not electronic; that is to say, they do not have tubes or transistors.

8·1 Basic motor control

An electric controller is a device, or group of devices, which governs electric power delivered to the apparatus to which it is connected. Controls range in complexity from a simple manual motor starter to an elaborate switchboard. Some basic control devices which may be built into controllers are described in the following paragraphs.

A *circuit breaker* has the function of interrupting the flow of power in a circuit under *normal* and *abnormal* conditions. Although it might be used as a power switch, its primary function is to protect the line against faults; hence it is designed for infrequent operation only. A *contactor* is a device for *repeatedly* establishing and interrupting an electric power circuit. Usually operated magnetically, it can be operated in line or be remotely governed by pilot devices or relays. The *motor-circuit switch* is used in motor branch circuits and may be used to interrupt maximum overload current or to remove the power supply from a starter and motor under a normal overload. A *relay* is a device made operative by varia-

(a)   (b)

Fig. 8·1   Two types of motor controllers. (a) Switch-type combination starter. (b) Synchronous motor starter. (Allis-Chalmers Mfg. Co.)

tions in the condition of a circuit and which, consequently, effects the operation of other devices in the same or other circuits. Although small itself (it operates on small currents and voltages), it can control other current-interrupting devices which may handle large currents and high voltages. As an example of a relay responsive to *overcurrent,* the variation in conditions would be a rise in current. The overcurrent causes the relay contacts to close, which in turn closes the breaker-trip circuit, which opens the breaker contacts.

Figure 8·2 shows seven typical methods of industrial control. Sometimes, when unexpected starting might prove dangerous to personnel, *undervoltage protection* is essential. (Unexpected starting of a motor-driven saw or shaper might be dangerous to the operator.) In Fig. 8·2d, pressing the *start* button, 1, energizes the contactor coil, M, which closes the contactor, 3. This closes the *interlock* circuit which parallels the start button contacts, and the line contactor is held "in" even though the start button contacts are open. When an undervoltage occurs the contactor opens, opening the interlock; but when the normal power returns, the line-contactor

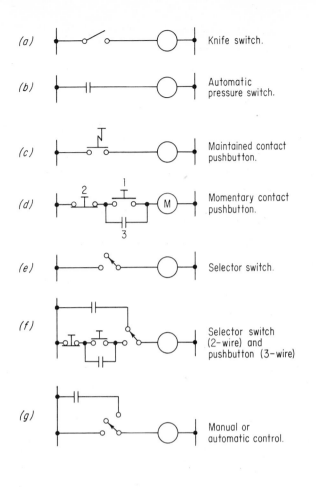

(a) Knife switch.

(b) Automatic pressure switch.

(c) Maintained contact pushbutton.

(d) Momentary contact pushbutton.

(e) Selector switch.

(f) Selector switch (2-wire) and pushbutton (3-wire)

(g) Manual or automatic control.

Fig. 8·2 Typical methods of industrial control. (d) and (f) are three-wire control. The others are two-wire control. (Allis-Chalmers Mfg. Co.)

coil is not energized until the start button is again closed. This normally open interlock is standard equipment on all magnetic starters. Button 2 is, of course, the *stop* button.

The other arrangements in Fig. 8·2 do not have undervoltage control except for Fig. 8·2f.

8·2 Functions of control

There are certain definite functions that are performed or governed by control. These are: (1) starting, (2) protection, (3) running, (4) speed regulation, and (5) stopping.

Reduced voltage starting—as an example of one type of control—reduces the current inrush during the starting period of a motor or machine. This eliminates or minimizes the shock of a quick start on the driven machine, the reduction of voltage in the line or system to which the machine is connected, and the dropping out of synchronism of synchronous motors on the line. Reduced voltage starters are available that provide smooth accelerated starting without serious drop in line voltage.

A large motor in a New England plant requires 56 sec starting time under normal load. An oil-well pump in Texas will suffer serious damage if its rotor locks and the motor is not tripped from the line in 20 sec. A conveyor drive motor in a Florida potash plant can withstand 25 percent overload for 30 min, but a compressor motor in Missouri may burn up in 3 min at 25 percent overload. Each of these motors must be protected from abnormal overload currents which creep in from time to time.

A momentary-contact push-button circuit with overload protection is shown in Fig. 8·3. An increase in current to the motor

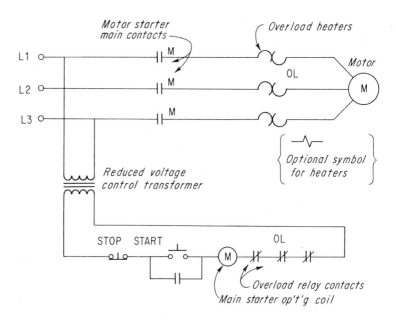

Fig. 8·3 *Elementary diagram of a push-button motor control circuit.* (*Plant Engineering Magazine.*)

is interpreted to be an overload by the overload heaters. The excessive current is translated into a temperature increase in the heaters which have a trip point (expressed in amperes). When this trip point is reached, the overload relays are opened and energy to the motor is stopped. It is good practice to have an overload heater on each incoming phase. Care must be exercised in selection of heaters because of such factors as ambient temperatures at the device.

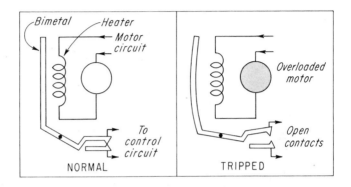

Fig. 8·4 Operation of a bimetallic-type overload relay. (*Plant Engineering Magazine.*)

8·3 Symbols for control diagrams

The starter diagram of Fig. 8·3 may appear a little strange to the reader because of the symbols which may appear to be unusual. This could be because of two factors: (1) different devices are often used in industrial controls than in computer and communication equipment, and (2) different, or alternative, symbols are often used for components that are commonly used.

A number of alternative symbols are in use in diagrams of this type. These include the contacts and heaters. Other symbols that may at first confuse the reader are the rectangular resistor symbol and some switches, such as the limit switch. In addition, distinctive types of drawings are often used. Two, for example, are the drum controller and the ladder diagram. Both will be explained in this chapter; the drum controller will be discussed next.

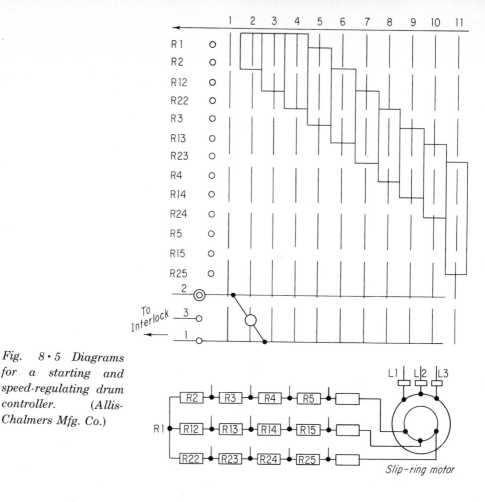

Fig. 8·5 Diagrams for a starting and speed-regulating drum controller. (Allis-Chalmers Mfg. Co.)

8·4 Drum controllers

The drum controller is very popular for control of shunt motors and as secondary control for wound-rotor motors. It is a sturdy device which lends itself well to good protective design. It has a cast cylindrical drum with contact segments which touch sliding finger contacts. A star-wheel notching device gives definite positioning on each contact point. The contact segments can be seen in the upper part of Fig. 8·5. This drawing should be thought of as a *development* of the cylindrical drum; in other words, the cylinder surface is laid out flat with the outside toward the reader.

When the motor is started, it is necessary to reduce the resistance as the speed increases. As the cylinder is rotated, resistances are cut out one by one until finally (at step 10) all resistance is out of the circuit. Step 11 is for speed regulation. In position 1, all resistance is in. In position 2, resistors $R1$-$R2$ are shorted out; in position 3, resistors $R1$-$R2$-$R12$ are shorted out. The lower diagram is a connection diagram showing how to connect resistor terminals to the corresponding fingers on the drum. The actual *interlock* with the primary magnetic contactor is not shown.

8·5 Constant-speed, electronics-controller circuit

The circuit of Fig. 8·6 automatically adjusts the motor voltage of a squirrel-cage motor in order to maintain a constant pre-

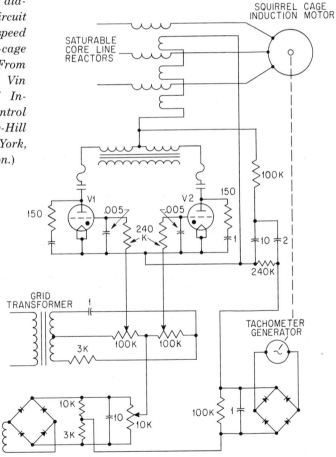

Fig. 8·6 A schematic diagram of a two-tube circuit for the adjustable-speed operation of a squirrel-cage induction motor. (From John Markus and Vin Zeluff, "Handbook of Industrial Electronic Control Circuits," McGraw-Hill Book Company, New York, 1956. Used by permission.)

selected speed, even if the load on the motor fluctuates or changes.

A preselected (reference) voltage, which corresponds to the desired speed, is set into the potentiometer, marked 10K at the center of the bottom of the diagram. Changes in motor speed appear in the a-c voltage generated by the tachometer generator, which is mechanically connected to the induction motor. This a-c voltage is rectified in the selenium bridge rectifier (beneath the tachometer) before going to the grids of the two thyratron (gas-filled triode) tubes. If the motor load decreases the motor speed will rise; the *difference* (sometimes called *error*) between the tachometer and reference voltages will cause the d-c output from the tubes to decrease. This reduces the saturation of the line reactors, which increases the portion of the incoming line voltage absorbed by them. Less voltage, then, goes to the motor, whose torque is reduced until it is just sufficient to handle the load at the desired constant speed. (Actually, the corrections are made before the motor changes its speed more than an infinitesimal amount.)

8·6 Symbols and identification

Symbols incorporated in this drawing are all standard ones. The symbol used for the tachometer generator is listed in ANSI Y32.2 and Mil Std 15 as being appropriate for oscillators, or a generalized a-c source. An alternate method would be to show a capital T within the circle or to place the abbreviation TACH beside the circle.

There are many other types of motor control circuits, both electronic and nonelectronic, for a-c and for d-c motors. It is hoped that the reader will have benefited from reading the foregoing material, although it is not complete.

One important factor in portraying these circuits is the use of appropriate designations for components. A set of standard device designations has been formulated by the Joint Industry Conference (JIC), which meets periodically to set up and modernize standards in this area of the electrical industry. Such a list appears in Table 8·1.

Table 8·1 Device designations

A	Accelerating contactor or relay	MTR	Motor
AM	Ammeter	MN	Manual
AU	Automatic	OL	Overload relay
BR	Brake relay	PB	Pushbutton
CAP	Capacitor	PL	Plug
CB	Circuit breaker	PLS	Plugging switch
CR	Control relay	PS	Pressure switch
CRA	Control relay automatic	R	Reverse
CRE	Control relay electronical energized	REC	Rectifier
CRH	Control relay manual (hand)	RECP	Receptacle
CRM	Control relay master	RES or R	Resistor
CT	Current transformer	RH	Rheostat
CV	Counter voltage	S	Switch
D	Down	SOC	Socket
DB	Dynamic braking contactor or relay	SOL	Solenoid
DISC	Disconnect switch	SCR	Series control relay
DS	Drum switch	SS	Selector switch
ET	Electron tube	SSW	Safety switch
F	Forward	T	Transformer
FLS	Flow switch	TB	Terminal board
FS	Float switch	THS	Thermostat switch
FTS	Foot switch	TR	Time delay relay
FU	Fuse	TVM	Tachometer voltmeter
GRD	Ground	Q	Transistor
IOL	Instantaneous overload	U	Up
LS	Limit switch	UCL	Unclamp
LT	Lamp	UV	Undervoltage
M	Motor starter	VM	Voltmeter
MB	Magnetic brake	VS	Vacuum switch
MC	Magnetic clutch	WM	Wattmeter
MCS	Motor circuit switch	X	Reactor

Relays:		Examples:	
	General use		CR 1CR 2CR
	Master		CRM
	Automatic		CRA
	Electronically energized		CRE 1CRE 2CRE
	Manual (hand)		CRH
	Latch		CRL 1CRL 2CRL
	Unlatch		CRU 1CRU 2CRU
	Timers		TR 1TR 2TR
	Overload relay		OL 1OL 2OL
	Motor starters		1M 2M etc.

SOURCE: JIC "Electrical Standards for Industrial Equipment."

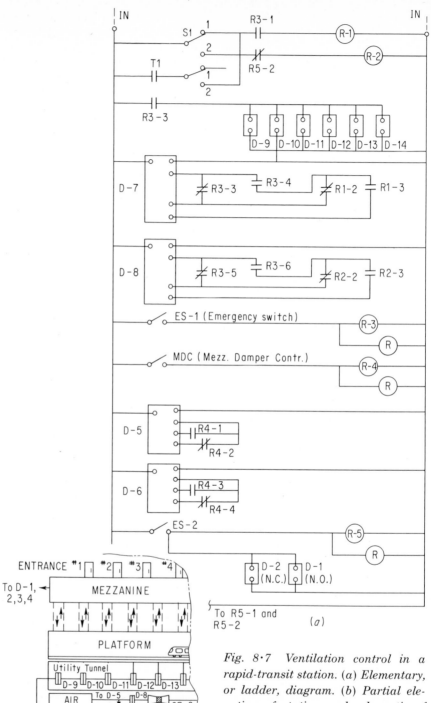

Fig. 8·7 Ventilation control in a rapid-transit station. (a) Elementary, or ladder, diagram. (b) Partial elevation of station and schematic of ventilation system. (Courtesy of S.F. Bay Area Rapid Transit District and Parsons-Brinckerhoff-Tudor-Bechtel, Engrg. Consultants.)

8·7 The ladder diagram

It is convenient to portray a circuit which causes, or coordinates, a series of events by means of a diagram such as that shown in Fig. 8·7. The signal flow is, in general, from top to bottom. The diagram shows the sequence of operations with regard to air flow in an urban mass-transit station that is mostly underground.

The sequence is as follows: Lead-lag switch $S1$ energizes relay $R1$ (or $R2$ if in No. 2 position), starting fan $SF2$. Outside thermostat $T1$ energizes relay $R2$ (or $R1$) to start $SF3$. When the second fan is on, damper motors $D9$ to $D14$ are actuated. Damper actuator $D8$ opens damper when $SF2$ runs. (The dampers are normally closed when fans are not running.) Damper actuator $D7$ opens damper when $SF3$ runs.

When emergency switch $ES1$ is energized, relay $R3$ opens dampers $D7$ and $D8$ and starts fans $SF2$ and $SF3$ in reverse, creating an exhaust situation (mode). Thermostat $T2$, in return air duct, controls damper actuator DM which is linked to dampers $D3$ and $D4$ in the ducts to outside air. A humidistat helps control heat and fresh-air intake. Emergency switch $ES2$ energizes relay $R5$ to close dampers $D1$ and $D3$ and open $D2$ and $D5$.

This diagram tells a story when it is printed alongside a diagram that shows the dampers, relays, and other parts involved. Without it, initial construction, daily operation, and maintenance would be difficult.

8·8 Automated machine tool

Figure 8·8 shows a numerically controlled drilling machine which is small enough to be economical on small runs as well as larger runs. It uses point-to-point numerical control, which is to say that it operates from X and Y dimensions relative to a zero point. A typical drilling project is shown in Fig. 8·9. The process sheet shows information necessary to produce a tape for the job shown in the drawing (Fig. 8·9a.) The program is then punched on a tape-punch machine, which looks like a typewriter, and the tape is fed into the machine. Some of the operations have the following symbols or abbreviations:

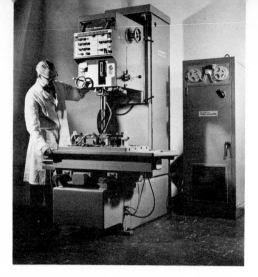

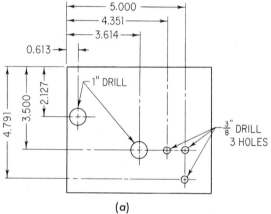

Fig. 8·8 Photo of a numerically controlled machine tool. (Pratt & Whitney Company, Inc.)

Fig. 8·9 Details of automatic drilling. (a) The drawing for a job to be drilled. (b) The process sheet for this job. (c) Standard 1-in. tape.

(a)

PROCESS SHEET

		COORDINATES				COMMENTS
MIS	TAB	LONGITUDINAL	TAB	TRANSVERSE	TAB	
	T	+ 00.000	T	+ 00.000	T	SET UP TO U.L. CORN. (EOB)
%	T	% + 00.613	T	− 02.127	T	DRILL 1" HOLES (EOB)
	T	+ 03.614	T	−03.500 (EOB)		
	T	&	T		T	TOOL CHANGE ³⁄₈"D (EOB)
& /	T	+ 04.351	T		T	DRILL ³⁄₈" HOLES (EOB)
	T	+05.000 (EOB)		−04.791 (EOB)		
			T			
		& /	T		T	TOOL CHANGE 1"D (EOB)

(b)

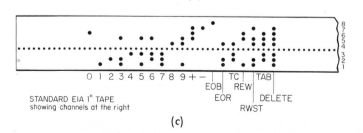

STANDARD EIA 1" TAPE
showing channels at the right

(c)

Stop rewind	%
Tool change light	&
Automatic rewind	/
TAB code (a spacer between dimensions on tape)	TAB
EOB code (end of block)[1]	EOB

The part to be drilled can be clamped anywhere on the table of the machine (eliminating the need for a jig) and either or both axes can be zeroed by a push button. The *tool change* code stops the machine cycling and advises the operator by means of a light to change the cutting tool. Aside from this stop, the machine drills the piece hole by hole automatically until the job is finished. Then the tape is rewound, and the machine is ready for the next part. Machines with multiple heads are even more automatic in that the operator does not have to change the drilling tools during the running cycle of the machine.

8·9 Drawings for an automated drilling machine

An elementary diagram of one part of an eight-head numerical drilling machine appears in Fig. 8·10. The term *elementary diagram* is used by the manufacturer. The author believes that a more appropriate title would be *block diagram*. However, because of the increasing complexity of many circuits, it is possible that the elementary diagram of the future may use blocks. This diagram is from one of several pages of such schematics. The signal flows are mainly from top to bottom, but there are also quite a few impulses going out the sides or coming in. Most of the blocks in this schematic represent NOR circuits. NOR circuit j k m, upper left, would function as follows:

j k m (inputs)	E11 n (output)
0 0 0	1
1 0 0	0
1 1 0	0
1 1 1	0
0 1 1	0
0 0 1	0
0 1 0	0

---

[1]This stops the passage of the tape through the tape reader, telling the machine to act on that information.

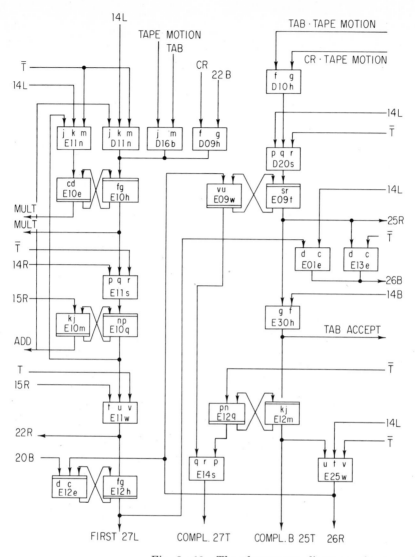

*Fig. 8·10 The elementary diagram of a translator opera-
tion selector of a numerical drilling machine. (Pratt &
Whitney Company, Inc.)*

The model, of which Fig. 8·10 is a circuit drawing, utilizes
mostly four solid-state circuit cards. Details of a typical logic
element, shown electrically in Fig. 8·10, are put on one
drawing, as shown in Fig. 8·11. It describes a printed circuit
card which has five NOR elements, each of which is represented by a rectangular block in Fig. 8·10 and also in the

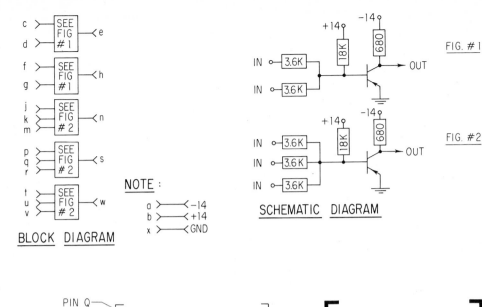

BLOCK DIAGRAM

NOTE :

a >——< −14
b >——< +14
x >——< GND

SCHEMATIC DIAGRAM

FIG. #1

FIG. #2

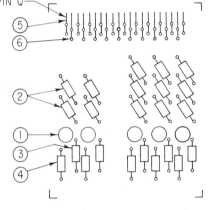

PIN Q

COMPONENT LAYOUT

CIRCUIT PATTERN

M−3042   U−38469

ITEM	QUAN.	P & W NO.
1	5	D007810−001
2	13	D007818−062
3	5	D007818−079
4	5	D007818−045
5	11	D007809−001
6	10	D007809−002

NOTE :
1. USE P.C. BOARD BLANK B−214090
2. USE ART WORK NO. W53515
3. COLOR CODE EDGE OF CARD YELLOW

Fig. 8 · 11 Details of a standard NOR *card for a numerically controlled drilling tool.* (*Pratt & Whitney Company, Inc.*)

block diagram, which is one of four drawings on Fig. 8·11. The other three drawings are (1) the printed circuit layout (pattern), (2) the physical arrangement of the components on the card, and (3) schematic drawings of the two NOR circuits, one having two inputs, the other having three inputs.

The reader can study the schematic diagram and the component layout diagram for a few minutes, after which he can probably identify the NOR circuit elements as they are physically located on the component layout. Then he can study the circuit pattern and component layout together to determine whether the printed circuit layout is on the opposite side of the component layout or not. Except for the dimensions of the printed circuit card, this drawing gives enough information for the production of this standard-card assembly. This type of solid-state-card assembly is used throughout the computer manufacturing industry, as well as in the automated machine tool industry.

Not all of the circuitry of a machine tool is electronic. Some is electromagnetic. The elementary diagram of the electromagnetic part of the numerical drilling machine is shown in Fig. 8·12. This drawing is typical of the older, more conventional, industrial control circuits. The signal flow is, in general, from top to bottom. The operation of various parts begins sequentially going down the diagram, as follows: After the operator has pushed the two start buttons, the *ready* light goes on, the *spindle* starts, then the *table* will traverse left or right, and so on until the cycle is completed. Some accessory equipment has been omitted from the diagram.

In this type of diagram, relay contacts and switches are generally connected to the left-hand vertical line (often called a bus), which is usually positive, and the relay coils and solenoids are connected to the right-hand line, which is usually neutral or negative. The connecting lines are usually numbered sequentially from upper left to lower right. The wire number is changed when the circuit passes through a component. (Note that wire No. 531 to the left of 1RES changes to No. 681 to the right of the resistor.) This also provides a convenient system for identifying components in a complex circuit. This type of drawing is sometimes called a *ladder diagram.* Each *level* may also be identified with numbers for locating parts of divided symbols. Such a system would be as shown in Fig.

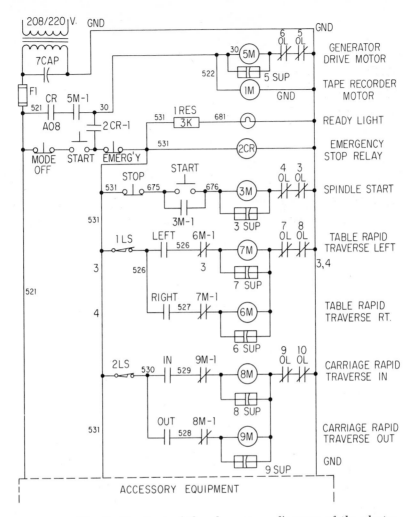

Fig. 8 · 12 Part of the elementary diagram of the electromagnetic portion of a numerical machine whose electronic
circuitry is described in Figs. 8 · 10 and 8 · 11.

8·12, in which the numbers 3 and 4 appear about halfway
down line No. 531. At *level 3* is suppressor 7 (7 *SUP*), and
at *level 4* is contact 7*M*·1. These are companion parts of
control relay 7*M*, which causes the table to start its leftward
traverse. 7*M*·1, however, has a timing function which readies
the system for the next step, in this case traversing to the
right. In this numbering system, the numbers 3 and 4 also

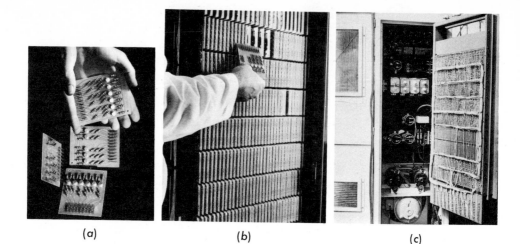

(a)　　　　　　　　　(b)　　　　　　　　　(c)

Fig. 8 · 13　Printed-circuit cards for a numerically controlled drilling machine. (a) Standard cards (NOR card in the center). (b) The arrangement (positioning) of the cards in the tape deck unit. (c) Some of the wiring associated with the circuit cards and other parts of the tape deck unit. (Pratt & Whitney Company, Inc.)

appear to the right of the control relay. The suppressors are combination resistors (220 ohms) and capacitors (1 microfarad) made as one unit, which serve to *hold in* (or lock in) the control relays until they are somehow de-energized. De-energization may be obtained by action of the overload relays or by pushing the stop, or emergency, buttons.

The drawings shown in Figs. 8 · 10 to 8 · 12 are typical of the many drawings required for the design, construction, and maintenance of automated machine tools. Where one drawing has been shown in this chapter, maybe twenty other similar drawings will be required to produce the machine. The simplest of them require about fifty pages of schematics to describe all the functions.

SUMMARY　The drawings in this chapter have been selected from hundreds of electrical systems that are employed in United States industry today. Although these are isolated examples, we have listed the main functions of *control* and seven typical methods of industrial control. Symbology in certain types of industrial electronics diagrams is sometimes different in appearance from electrical diagrams previously shown in this book. In some cases, nonstandard symbols are used. In other cases, the types of devices used, such as relays and contacts, have their own special symbols. Logic diagrams are widely used where automated circuitry makes use of logic circuits. Many of these circuits are put on printed-circuit cards which can be removed and replaced easily. Identification of components, including relays, switches, and logic circuits becomes important because of the many devices used in these circuits. Standard identifications as well as standard symbols should be employed.

QUESTIONS

8·1 Name four of the main functions of motor control.

8·2 What main function is found in practically all controllers?

8·3 What establishes whether a controller or control system is electronic or not?

8·4 What is the difference between a contactor and a circuit breaker in so far as their functions are concerned?

8·5 What are the basic building blocks of an ultrasonic system?

8·6 List four applications of ultrasonics that require cavitational action, as well as vibrational energy.

8·7 What is the dividing line between ultrasonic and sonic energy?

8·8 What does EOB stand for in the code used for an automatic machine tool? Explain.

8·9 What is the input of a numerical drilling machine, and what is the output?

8·10 What types of drawings are required for the construction of an automated drilling machine?

8·11 For what words do the following abbreviations stand: PB, 5POT, CR1, LS, OL, TR?

8·12 What standard reference is widely used for determining what symbols are to be used in industrial control schematic drawings?

8·13 What standard reference may be used to determine what standard identifications are to be used in industrial control drawings?

8·14 Sketch three nonstandard symbols that have been used in some industrial electronics elementary diagrams.

8·15 What would a long line consisting of dashes between a motor symbol and a tachometer represent?

8·16 What is an advantage of using the rectangular resistor symbol rather than the zigzag symbol? A disadvantage?

8·17 Sketch two approved symbols for a relay coil.

8·18 If A and B are the inputs of a NOR circuit, what state is the output C in if A is positive and B is negative?

8·19 If R, S, and T are the inputs of a NAND circuit, in what state is the output U if R and S are positive and T is negative?

8·20 What type of tube is a thyratron tube (one is shown in one of the drawings in this chapter)? Is there a transistor that can perform the same function?

8·21 What is meant by reference voltage?

8·22 What is meant by *difference* or *error?*

8·23 What does a transducer do? Name two examples of transducers.

PROBLEMS 8·1 Laser problem. Light amplification by stimulated emission of radiation has four basic units: (1) laser pump, (2) laser material, (3) laser optical cavity, and (4) optical focusing system. Figure 8·14 shows a schematic diagram of a laser system. The laser material may be of crystal, glass, semiconductor material, or gas. Make a set of drawings including:
a. The diagram of Fig. 1·14
b. The laser pump pulse-forming network (Fig. 8·14a)
c. A wideband optical system diagram using a laser beam (Figure 8·14b.)

A laser beam can handle extremely high [such as gigacycles (10^9 CPS)] frequencies with high-directional capabilities. Some other industrial uses for lasers are machining, welding, seismography, and continental drift detection. Use 8½ × 11 paper.

8·2 Draw the diagram of a step controller shown in Fig. 8·15. Use a push-button switch symbol instead

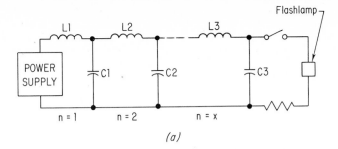

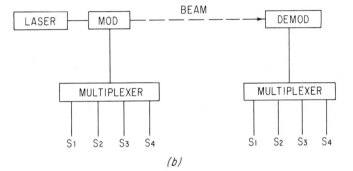

Fig. 8·14 (Prob. 8·1) Laser details. (a) Pump pulse-forming network. (b) Optical communication system using a laser.

Fig. 8·15 (Prob. 8·2) A ladder diagram for an electric step controller.

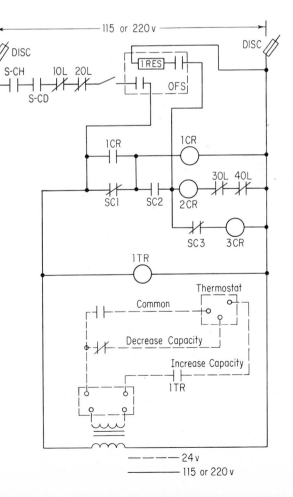

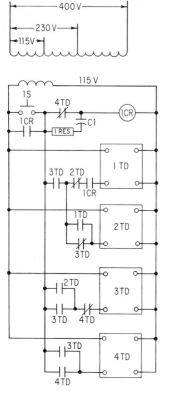

Fig. 8·16 (Prob. 8·3) A ladder diagram for a sequence timer.

of the knife switch shown. The following abbreviations are used: S-CH—chilled water pump starter; S-CD—condensed water pump starter; OFS—oil failure switch; ICR—recycle relay; SC1—step controller switch; 2CR—compressor motor starter; 3 CR—capacity, reduction solenoid, and 1TR interval timer. Use 8½ × 11 paper.

8·3 Make a drawing of the ladder diagram for the sequence timer shown in Fig. 8·16. The following changes might be in order: Add a stop push-button switch and selector switch to the appropriate part of the diagram. Add overload protection for the delay-period circuits and other places you deem advisable. Use 8½ × 11 or 11 × 17 paper.

8·4 Adjustable speed drives for d-c motors are of four types: (1) motor generator, (2) electron tube, (3) magnetic amplifier, and (4) controlled rectifier. Although still around, the tube and magnetic-amplifier drives are obsolescent. Figure 8·17 shows in diagrammatic form the two types that are still popular. Draw these two diagrams and label all parts. (3φ means 3-phase, SCR means silicon controlled rectifier.) Use 8½ × 11 paper.

8·5 Design a controls system for a system having several motors or circuits, and make a complete drawing for this controls system. One example might be an air conditioning system of a large building which employs three compressors. When the outside temperature rises a certain amount, the first compressor will start operating. When it is another 10° hotter, the second compressor comes on, etc. Show overload protection.

8·6 Figure 8·18 includes drawings for a translator circuit, which has many NAND circuits. Make a complete set of drawings for the translator circuit, including the truth table, which should be completed. Use 8½ × 11 or 11 × 17 paper.

8·7 Figure 8·19 shows dimensions of the FLAT PAK package. These packages come in different sizes and are available in many different circuits. Typical circuits are video amplifier, binary counter, dual NAND

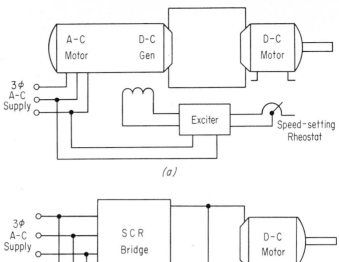

(a)

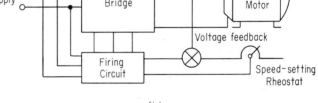

(b)

Fig. 8·17 (Prob. 8·4) Two types of adjustable-speed drives. (a) Motor-generator. (b) Controlled rectifier drive.

Fig. 8·18 (Prob. 8·6) Drawings for a translater circuit.

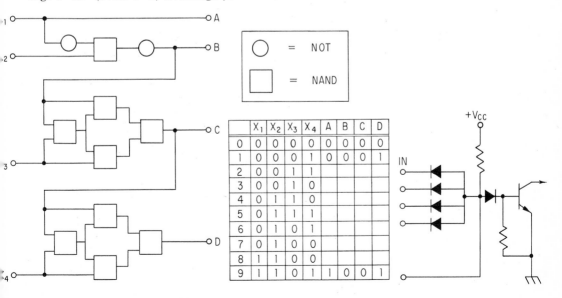

	X_1	X_2	X_3	X_4	A	B	C	D
0	0	0	0	0	0	0	0	0
1	0	0	0	1	0	0	0	1
2	0	0	1	1				
3	0	0	1	0				
4	0	1	1	0				
5	0	1	1	1				
6	0	1	0	1				
7	0	1	0	0				
8	1	1	0	0				
9	1	1	0	1	1	0	0	1

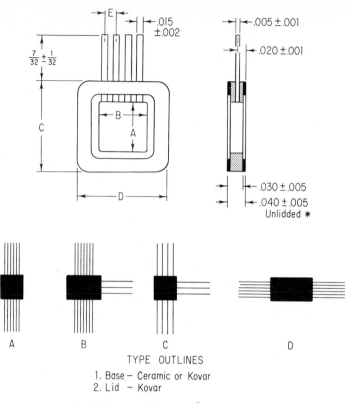

TYPE OUTLINES
1. Base — Ceramic or Kovar
2. Lid — Kovar

* Lid increases this by .010″

Fig. 8·19 (Prob. 8·7) Details for functional block packages. (Westinghouse Electric Corp.)

gate, and *RS* flip-flop. Make a plate which contains all the drawing and dimensions shown in Fig. 8·19, plus the table shown below. Use 8½ × 11 paper.

package type	A	B	C	D	E
A	.120	.120	¼ ± ⅟₃₂	¼ ± ⅟₃₂	.050 ±.005
B	.120	.200	¼ ± ⅟₃₂	⅜ ± ⅟₃₂	.050 ±.005
C	.200	.200	⅜ ± ⅟₃₂	⅜ ± ⅟₃₂	.050 ±.005
D	.200	.600	⅜ ± ⅟₃₂	¾ ± ⅟₃₂	.050 ±.005

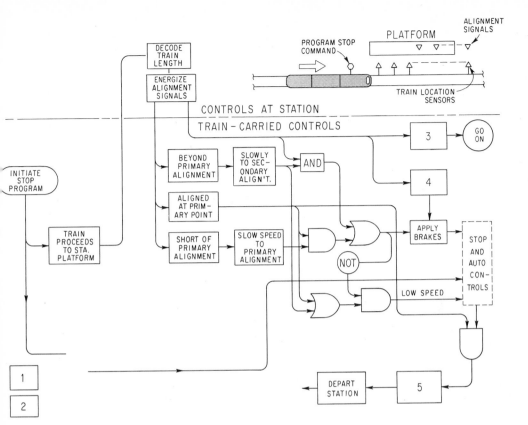

Fig. 8·20 (Prob. 8·8) Functional diagram of rapid-transit
train stopping program. (From S.F. BART District.)

8·8   Figure 8·20 is a diagram of the train-stopping control
program for a rapid-transit system. Draw this functional diagram on an 11 × 17 sheet. In order to
complete it you must show the following: Box 1, Adjust
stopping profiles for this train length; Box 2, Adjust
stopping profile for this station. (These two functions
must be considered as input along with the box, Initi-
ate stop program, to an AND function in the lower
left sector.) Also: Box 3, Beyond secondary alignment
point, signal lost; Box 4, Aligned at secondary point,
signal received; and Box 5, Aligned and stopped, open
doors command. Also, put a program end command
at the end of the function diagram. The train and
platform detail may be added as shown. Be consistent
in the type of logical symbols you use for the logic
functions.

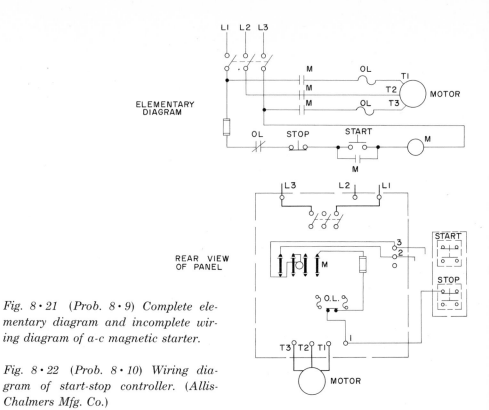

ELEMENTARY
DIAGRAM

REAR VIEW
OF PANEL

Fig. 8·21 (Prob. 8·9) Complete elementary diagram and incomplete wiring diagram of a-c magnetic starter.

Fig. 8·22 (Prob. 8·10) Wiring diagram of start-stop controller. (Allis-Chalmers Mfg. Co.)

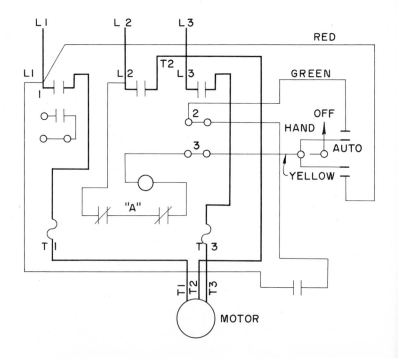

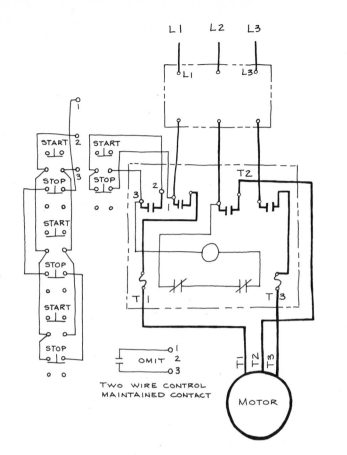

L1 L2 L3

L1 L3

START 2
STOP 3

START
STOP

START
STOP

START
STOP

OMIT 1
        2
        3

TWO WIRE CONTROL
MAINTAINED CONTACT

T2

3   2

T1   T3

T1 T2 T3

MOTOR

*Fig. 8·23 (Prob. 8·11)
Sketch of a diagram of
a three-phase three-pole
starter.*

8·9   Figure 8·21 shows the complete elementary diagram
       and the incomplete panel-wiring diagram of an a-c
       magnetic across-the-line starter with three-wire con-
       trol from a remote push-button station. Redraw the
       two diagrams, completing the panel diagram. Use
       8½ × 11 paper.

8·10   Redraw the wiring diagram for the three-pole starter
       shown in Fig. 8·22. Add the elementary diagram to
       your drawing if your instructor so indicates. Add the
       missing terminal for AUTO mode to the switch sym-
       bol. Use 8½ × 11 paper.

8·11   Make an instrument drawing of the freehand sketch
       shown in Fig. 8·23. Use 8½ × 11 paper.

Chapter 9
Drawings for the electric power field

Drawings that are made for electric power installations can be as complex as those made in the various electronics areas. In fact, a complete set of drawings for a typical thermal generating station just about covers the entire field of electrical drawing. Included in such a set, for instance, would be schematic (elementary) diagrams, pictorial drawings, one-line circuit drawings, building wiring drawings, connection diagrams, and logic diagrams, as well as plans, elevations, sections, and details of structures and equipment.

Figure 9·2 shows a typical drawing arrangement that is used for the design of a power station. It includes a sche-

Fig. 9·1 Photo of a substation. (Kansas Power & Light Co.)

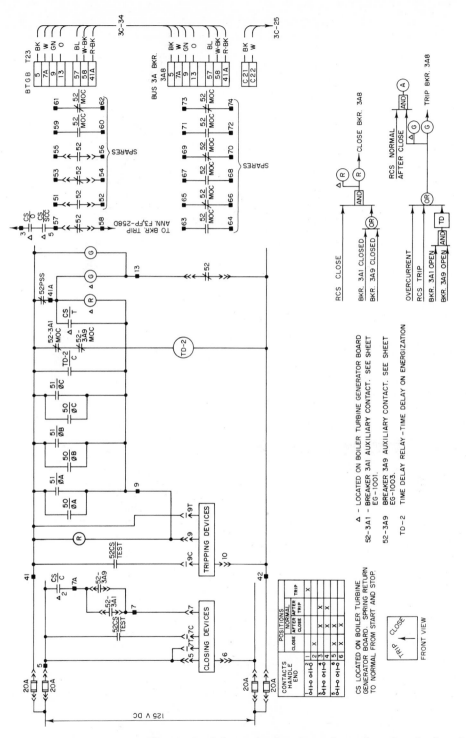

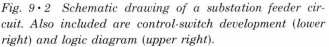

Fig. 9·2 Schematic drawing of a substation feeder circuit. Also included are control-switch development (lower right) and logic diagram (upper right).

matic diagram of the control system for a circuit, plus a diagram showing how the control switch functions, and a logic diagram of the system. Some of the abbreviations which may not be familiar to the reader are: CS, *control switch;* BKR, *circuit breaker;* BTGB, *boiler turbine generator board;* SCC, *station control cabinet;* PSS, *permissive selector switch;* RCS, *remote control switch;* MOC, *mechanically operated contact;* and φA, *phase, A.*

9·1 Logic diagrams

Logic diagrams provide an effective means of communication among persons involved in design, construction, operation, and maintenance of electric generating stations. The five basic symbols are shown and explained in Fig. 9·3. Also shown are

Fig. 9·3 Basic logic symbols. The time-delay symbol numbers give the range of adjustment desired in seconds.

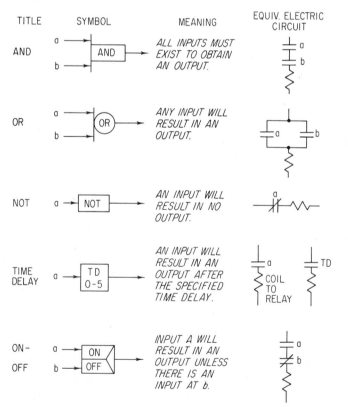

TITLE	SYMBOL	MEANING	EQUIV. ELECTRIC CIRCUIT
AND	a → AND → b	ALL INPUTS MUST EXIST TO OBTAIN AN OUTPUT.	a, b
OR	a → OR → b	ANY INPUT WILL RESULT IN AN OUTPUT.	a, b
NOT	a → NOT →	AN INPUT WILL RESULT IN NO OUTPUT.	a
TIME DELAY	a → TD 0-5 →	AN INPUT WILL RESULT IN AN OUTPUT AFTER THE SPECIFIED TIME DELAY.	a COIL TO RELAY, TD
ON– OFF	a → ON / b → OFF →	INPUT A WILL RESULT IN AN OUTPUT UNLESS THERE IS AN INPUT AT b.	a, b

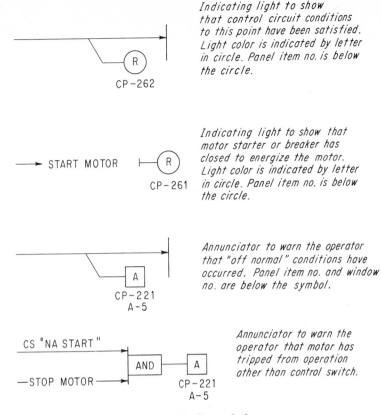

CP-262

*Indicating light to show
that control circuit conditions
to this point have been satisfied.
Light color is indicated by letter
in circle. Panel item no. is below
the circle.*

→ START MOTOR ⊢ R

CP-261

*Indicating light to show that
motor starter or breaker has
closed to energize the motor.
Light color is indicated by letter
in circle. Panel item no. is below
the circle.*

A

CP-221
A-5

*Annunciator to warn the operator
that "off normal" conditions have
occurred. Panel item no. and window
no. are below the symbol.*

CS "NA START"

AND — A

—STOP MOTOR →

CP-221
A-5

*Annunciator to warn the
operator that motor has
tripped from operation
other than control switch.*

Fig. 9·4 Supplementary logic symbols.

typical equivalent electrical circuits. Some supplementary logic symbols are shown in Fig. 9·4. These symbols convey necessary information used by the electrical control designer, and with their associated descriptive information (panel item, numbers, etc.) aid the overall project design coordination.

The symbols shown do not agree entirely with any one manufacturer's standard, but they are similar to the NEMA (National Electrical Manufacturers' Association) standard. No particular device is represented by one of these symbols. This type of diagram, therefore, leaves the design of the circuit (including arrangement and selection of devices) up to the electrical control designer.

9·2 Relation of logic and schematic diagrams

Figure 9·5 shows the logic diagram(s) and schematic diagram of a system that provides correct operation and supervision of

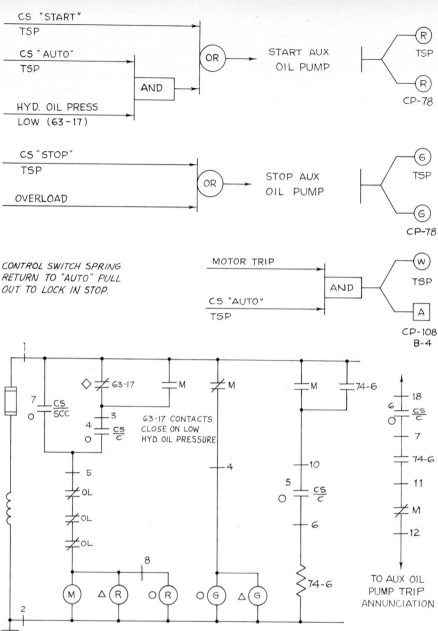

CS "START"
TSP

CS "AUTO"
TSP

HYD. OIL PRESS
LOW (63-17)

AND

OR → START AUX OIL PUMP

R TSP
R CP-78

CS "STOP"
TSP

OVERLOAD

OR → STOP AUX OIL PUMP

G TSP
G CP-78

CONTROL SWITCH SPRING
RETURN TO "AUTO" PULL
OUT TO LOCK IN STOP.

MOTOR TRIP

CS "AUTO"
TSP

AND

W TSP
A CP-108 B-4

7 CS/SCC
4 CS/C
3
63-17

63-17 CONTACTS
CLOSE ON LOW
HYD. OIL PRESSURE

M   M   M   74-6

5
OL
OL
OL

8

M △R ○R ○G △G

4

10
5 CS/C
6

74-6

18
6 CS/C
7
74-6
11
M
12

TO AUX OIL
PUMP TRIP
ANNUNCIATION

CONTROL SWITCH DEVELOPMENT DELETED

△ LOCATED ON CONTROL PANEL
○ LOCATED ON TURBINE START UP PANEL
◇ LOCATED LOCALLY

Fig. 9·5 Example of a logic and a schematic diagram for an auxiliary oil-pump control system. (Black & Veatch, Consulting Engineers.)

an auxiliary oil pump. The logic diagram is interpreted as follows:

If the control switch is in the start position or in the automatic position and *the turbine hydraulic-oil pressure is low, the pump starts. When the starting device is actuated or closed, a red light will go "on" on the turbine start-up (TSP) panel and on the control (CP) panel. (The control switch is on the TSP panel, incidentally.)*

If the control switch is in the stop position or *if there is a motor overload, the pump stops. In this case a green light will be energized on the CP and TSP panels.*

If the motor had started and then tripped for some reason other than a control-switch action, an alarm will sound on the visual annunciator (item CP-108, window B-4) and a white light will be energized on the TSP panel.

Certain conventional functions or actions are understood in the reading of these diagrams.

1 Unless otherwise indicated, control switches, push buttons, and selector switches are understood to be spring-return to normal.
2 Electrical interlock devices such as cell interlocks are not shown but are understood to be present in accordance with the clients' or office standards.
3 "Motor trip" (as contrasted to "stop") is understood to mean that the motor has stopped for some reason other than by a control-switch action.

The schematic diagram which applies to the logic diagram in Fig. 9·5 explains in detail all the items necessary for accurate electrical construction, operation, and also becomes a valuable maintenance tool. The control-switch development normally shown on this type of drawing has been deleted for simplicity.

It is possible for the system designer to dispense with schematic diagrams and to employ only logic diagrams in his communication with the fabricator of the control systems. However, according to one designer, the following reasons justify the making of schematic diagrams for most systems:

1 To show a complete system with all desired electrical features

2 To control the selection of alternates in the circuitry and components to be used by the manufacturer

3 To permit early determination and retain control of interconnecting cable requirements

9·3 The one-line diagram

One of the first drawings made in a set of drawings for the design of a large project is a one-line diagram of the entire plant or system. Such a drawing shows by means of single lines and symbols the major equipment, switching devices, and connecting circuits of a plant or system.

Figure 9·6 is a good example of a one-line diagram. The single line running down through the figure actually represents three lines in this three-phase system. The power path can easily be traced from the shielded ACSR (aluminum cable, steel reinforced) downward past a grounded lightning arrester, through a GOAB (trade name) disconnect switch, fused disconnect switches, and a step-down transformer. It continues on down through an oil circuit breaker, and auxiliary equipment. However, part of the electricity is sent through potential transformers where it is stepped down to a small voltage for metering. In addition, current transformers near the OCB are used for current-measuring meters, none of which is shown in the drawing, although each is referred to in a note. From here, the power goes through a disconnect switch, past another lightning arrester and out through a 12.47 KV FCWD (type F Copperweld) line.

Salient features which should be observed in the making of one-line drawings, as shown in Fig. 9·6, are:

1 Use of standard symbols, abbreviations, and designations

2 Highest voltage lines placed at the top or left of the drawing, if possible, with successively lower voltages placed downward or to the right

3 Main circuits drawn in the most direct and logical sequence

4 Lines between symbols drawn either vertically or horizontally with minimum crossing of lines

5 Generous spacing to avoid crowding symbols and *notes*

6 When pertinent, information should be included on rating

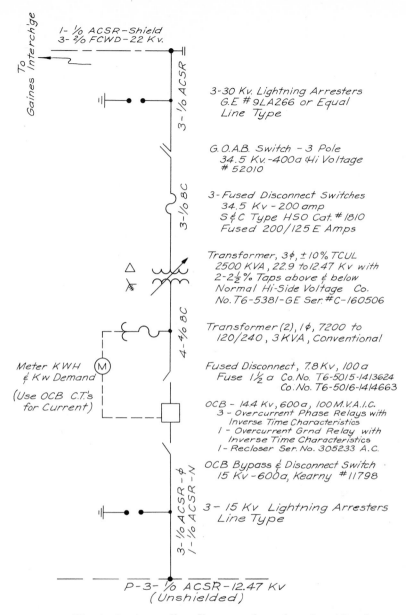

To Gaines Interch'ge
1- 1/0 ACSR-Shield
3- 2/0 FCWD-22 Kv.
3-1/0 ACSR
3-30 Kv. Lightning Arresters
G.E #9LA266 or Equal
Line Type
G.O.A.B. Switch - 3 Pole
34.5 Kv.-400a Hi Voltage
#52010
3-1/0 8C
3-Fused Disconnect Switches
34.5 Kv - 200 amp
S&C Type HSO Cat. #1810
Fused 200/125 E Amps
Transformer, 3φ, ±10% TCUL
2500 KVA, 22.9 to 12.47 Kv with
2-2½% Taps above & below
Normal Hi-Side Voltage Co.
No. T6-5381-GE Ser. #C-160506
4-4/0 8C
Transformer (2), 1φ, 7200 to
120/240, 3 KVA, Conventional
Meter KWH
& Kw Demand
(Use OCB C.T.'s
for Current)
M
Fused Disconnect, 7.8 Kv, 100 a
Fuse 1½ a Co. No. T6-5015-1413624
Co. No. T6-5016-1414663
OCB - 14.4 Kv, 600 a, 100 M.V.A.I.C.
3 - Overcurrent Phase Relays with
Inverse Time Characteristics
1 - Overcurrent Grnd Relay with
Inverse Time Characteristics
1 - Recloser Ser. No. 305233 A.C.
OCB Bypass & Disconnect Switch
15 Kv - 600a, Kearny #11798
3-1/0 ACSR-φ
1-1/0 ACSR-N
3 - 15 Kv Lightning Arresters
Line Type
P-3-1/0 ACSR-12.47 Kv
(Unshielded)

Fig. 9 • 6 A one-line diagram of a substation. (*Southwestern Public Service Company.*)

of equipment, feeder lines and transformers, vector relationship of equipment, ground or neutral connection of power circuits, protective measures, types of switches and circuit breakers, and instrumentation and metering.

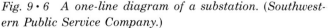

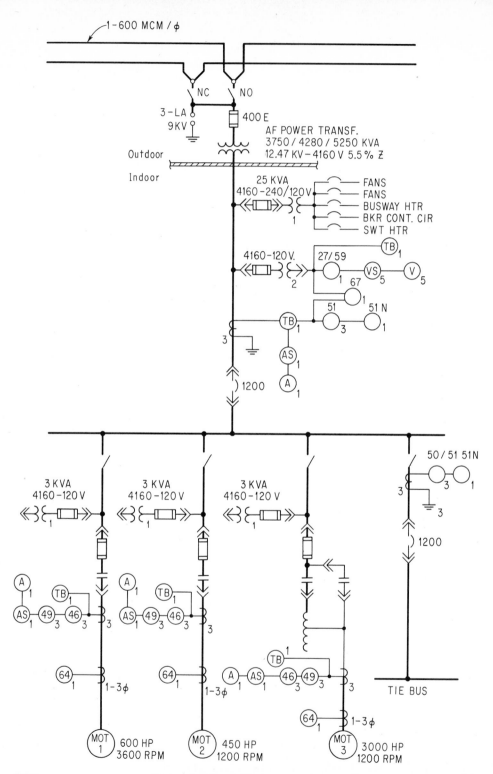

Fig. 9·7 Part of a one-line diagram for an industrial plant.
(Western Electric Co.)

If the above points are observed in the making of a one-line diagram, the drawing will impart a clear picture of the system so represented. The one-line drawing, which is used for many purposes, is a distinctive drawing peculiar to the electrical field and as such has made a great contribution to the industry.

9·4 Other one-line drawings

Approximately one-half of a one-line diagram for a part of a large electronics manufacturing plant is shown in Fig. 9·7. This diagram shows the electric energy coming from the power company after having been reduced to 12.47 KV. A small substation on plant property reduces this supply to 4.16 KV by means of an Askeril-filled (AF) transformer. The 600-MCM conductor has an area equal to 600,000 circular mils. A circular mil is a unit used in specifying the cross-sectional area of round conductors. It is equal to the area of a circle whose diameter is 1 mil (0.001 in.).

The path of the energy can be traced to the motors at the bottom. These motors turn compressors and chillers for the plant air-conditioning. To facilitate reading the diagram a legend showing the abbreviations was printed on the diagram. It is reproduced below:

A	Ammeter	50/51	Overcurrent (instantaneous and time) relay
AS	Ammeter switch		
V	Voltmeter		
VS	Voltmeter switch	5IN	Overcurrent (ground) relay
TB	Test block		
27/59	Under/over voltage relay	64	Ground-fault relay
		67	Reverse-power relay
46	Phase-failure relay	87	Differential-protective relay
49	Overcurrent-over-load relay		

To the right of and below most of the symbols is a number. This number indicates how many of the devices are represented by the symbol. A 3, for instance, means that there will be three devices (one on each line) at that location.

In accordance with ANSI Y14.15, "Electrical and Electronics Diagrams," the primary conductors have been shown with heavy lines. Some other details recommended by this standard are:

1 Winding connection symbols should be shown for all power equipment.

2 Neutral and ground connections should be shown for all grounded equipment.

3 Rating and type of load, when available, should be shown for each feeder circuit.

4 Current transformers should show the ratio of transformers and the ampere ratio. (Example: 3 — 600/5.)

5 Industrial-control single-line diagrams may omit equipment ratings when they are used as standard drawings applying to more than one rating.

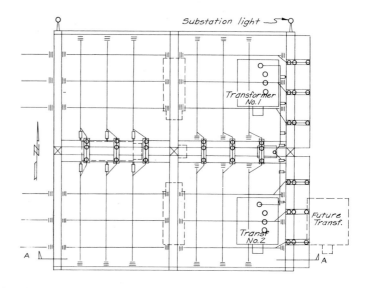

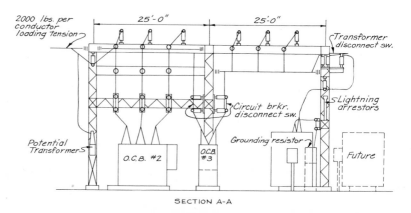

SECTION A-A

Fig. 9·8 Part of general layout of a substation. (Black & Veatch, Consulting Engineers.)

Preparation of specifications is an important part of the engineer's job. The person who makes electrical drawings can expect to do much work on the specifications, which often consist of several hundred pages for a single installation.

9·5 General arrangement

Another type of drawing which usually accompanies circuit drawings previously described in this chapter is the *general layout,* or *general arrangement,* of the plant or station. Such a drawing may contain a plot plan, detail plan of the installation, sections, and a small, simple, one-line diagram. Figure 9·8 shows only the detail plan, and one of three sections which are included on the same sheet.

These drawings are very simple and neat, and are good examples of orthographic projection as it applies to the electrical industry. Occasionally an isometric, or some other type of pictorial view, is used either as a supplement to, or a substitute for, the views shown here.

9·6 Other circuit drawings

Other circuit drawings may be necessary or desirable, depending on the arrangement, equipment, and other features of the installation being described. Various control circuit wiring diagrams are customarily shown for example. Similar drawings for power and other circuits are also in common use. Imagination and ingenuity are often called upon in order to put down on paper information about a particular system. This is illustrated by the drawing showing thermocouple locations (see Fig. 9·9). Here, because of the importance of having thermocouples installed at the correct heights, the elevation of each device is shown beside its location. In this case, the elevations refer to an arbitrary elevation of 100.00 feet which was assigned to the ground floor level. Accompanying the diagram, and on the same sheet (but omitted in Fig. 9·9), is a schedule in which each of the 70-odd thermocouples is described in this manner:

no.	*location*
1	Super heater outlet
2	Reheater outlet
3	Main steam to turbine

and so on for each thermocouple.

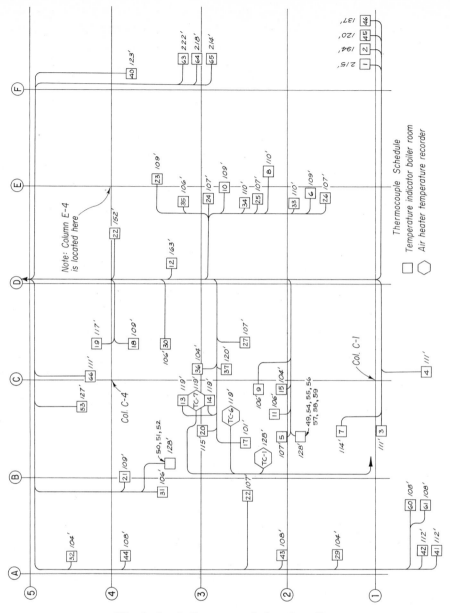

Fig. 9·9   A thermocouple location diagram.

This is a very simple drawing, yet its very simplicity and uncrowded appearance make it very effective. Too much emphasis cannot be given to the planning of uncrowded drawings. Putting too much information on a drawing, in an effort to save a little blueprint paper, may result in a drawing that is difficult and time-consuming to read. This is false economy.

9·7 Details

No set of drawings covering a power installation of any size is complete without detail drawings. There may be several large sheets filled with small detail drawings. These may be roughly divided into two categories: (1) special, and (2) standard details.

Special details, which cover the particular project being delineated, may include many different items, installations, or interconnections.

Examples of these drawings appear in Fig. 9·10. Two views showing the arrangement of trays are given (see Fig. 9·10a). These are rectangular metal ducts in which conduits are run throughout the structure. The conduits themselves are usually not shown, but their tray placement can be determined elsewhere, such as in the circuit routing designation in the specifications referred to in Sec. 9·4. Figure 9·10b shows a section of a *duct bank,* which is an underground encasement having six banks of 3-in. Korduct in which cables are run. Korduct is a thin-walled pipe used for encasement in concrete. Transite pipe is used for direct burial in the ground as are ducts of fibre and plastic.

These are just three of many types of details that are encountered in drawings for the electric power industry. Standard details, which may be used for more than one project, are usually drawn just once, but reproduced many times. Sometimes they are reduced to 8½ × 11-in. size, then bound in the printed specifications.

9·8 General trends

The number of control circuits in generating station construction is so large that drawing them on one-line diagrams has been largely abandoned in favor of logic diagrams. Similarly, the practice of drawing circuits on electrical plans has given way to showing raceway numbers (trays, conduits, etc.) only. All details are organized in the raceway and conduit schedules and by interconnection diagrams or schematics. Figure 9·10a shows tray details in elevation; similar drawings are used to show tray and raceway details in plan (top) views.

There has been a tremendous expansion in the use of static elements, including semiconductors and magnetic amplifiers in the electric power field. A partial list is as follows:

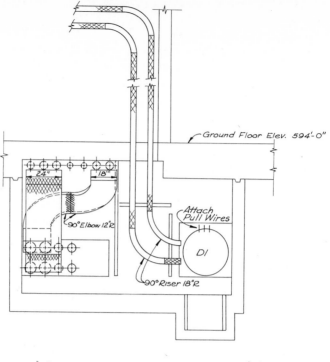

Ground Floor Elev. 594'-0"

24" 18"

Attach
Pull Wires

90° Elbow 12"R

90° Riser 18"R

D1

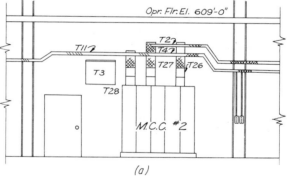

Opr. Flr. El. 609'-0"

T11   T2
       T4

T3   T27 T26

T28

M.C.C. #2

(a)

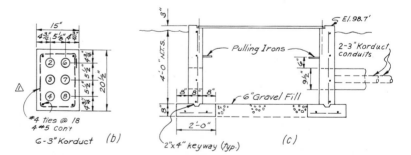

15"

4¾" 5⅛" 4¾"

2   6
3   7
4   8

4¾" 5½" 5½" 4¾"   20½"

#4 ties @ 18
4 #5 con't

6-3" Korduct   (b)

3"

El. 98.7'

4'-0" N.T.S.

Pulling Irons

2-3" Korduct
conduits

6"

9½"

6" Gravel Fill

8" 8" 8"

8"

2'-0"

2"x4" keyway (Typ.)

(c)

Fig. 9·10   *Special details that are part of a set of drawings for a power generating station.* (a) *Tray details.* (b) *Duct bank.* (c) *Pull box.* (Black & Veatch, Consulting Engineers.)

1 Static exciters, which are a combination of semiconductors and magnetic amplifiers, to electric utility generators, with capabilities up to 100,000 kw

2 Similar devices used as battery chargers and industrial rectifiers

3 Silicon controlled rectifiers used as power inverters (a-c to d-c and a-c to a-c) and as static switches with ratings of 50 to 100 amp

4 Protective relays of static elements which largely automate large circuit breakers

5 Systems for controlling power using more static devices

The ultimate in this direction is represented by "on-line" computers being used in power stations and factories.

SUMMARY The explanations and drawings in this chapter do not, by any means, give a complete picture of drawing which is performed in the power field. However, it is hoped that the drawings are typical enough and the coverage broad enough that the reader will have a fair knowledge of what type of drawing is done in this area.

In general, all types of electrical drawings are prepared. The one-line diagram is almost unique to the power field. Detail drawings are necessary, and many of these are structural-type drawings. Control circuits and power circuits for large installations have to be diagrammed separately. Volumes of printed specifications accompany sets of drawings. Engineers and draftsmen engaged in the preparation of such drawings may expect to participate in the preparation of these specifications.

QUESTIONS 9·1 Name three types of drawings that are often used in the electric power field.

9·2 How can one indicate how more than one relay is used at a location on a one-line diagram?

9·3 In addition to electrical drawings, what other types of drawings are associated with the complete description of a generator station?

9·4 Name five different kinds of relays that might be shown on a one-line diagram for an electrical substation.

9·5 Describe briefly what the following abbreviations stand for: C192, MH5, OCB, FCWD, R221, MCM.

9·6   At which end of a one-line diagram are the high-voltage lines and equipment placed?

9·7   Show by means of sketches how you would indicate a wye-delta transformer connection on a one-line diagram.

9·8   In a set of drawings used for the construction of a generator station, which drawings might be reasonably expected to utilize the "highway" principle discussed in Chap. 4?

9·9   In a set of drawings completely describing a generator station, of what particular installations would you expect to find sectional views?

9·10   Describe briefly two types of meters that may be found in drawings which are typical of the electric power field.

9·11   In addition to a meter, what other device is necessary in order that the KWH or KW demand of a high-voltage line can be measured?

9·12   What do the abbreviations CT, PT, OCB, and VS stand for?

9·13   What are the five main functions as used in logic diagrams for the electric power field?

9·14   What type of circuits can be appropriately shown or described by means of logic diagrams?

9·15   What is the distinction between the words *stop* and *trip* in so far as motor action is concerned?

9·16   What is understood about push buttons and control switches in the reading of logic diagrams shown in this chapter?

9·17   How does an overload relay work, as described in this book? Can you think of another type of design that would cause a current to stop flowing?

9·18   What is meant by *trays* in electric power construction? By duct banks? By raceways?

9·19   What are some of the trends in this field, in regard to the type of equipment that is being used?

9·20   Show by means of sketches two shapes of logic symbols that have been used for each of several common logic functions.

PROBLEMS   9·1   Complete the one-line diagram of the substation shown in Fig. 9·11 by adding the missing symbols. Place suitable identification adjacent to each symbol.

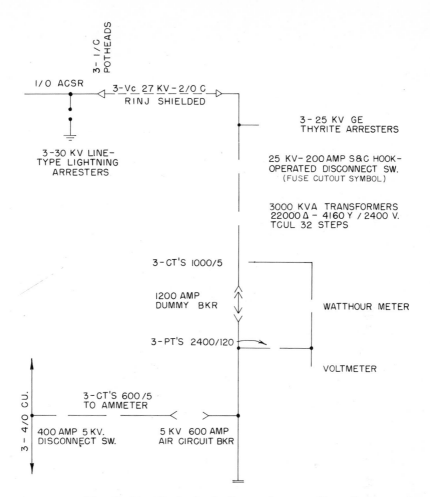

Fig. 9·11 *(Prob.* 9·1) *Incomplete one-line diagram of substation.*

9·2   Starting at the top and going down, draw the symbols in the following list (a.–k.) on a vertical line (except where noted). Place identification of each symbol at the right of symbol on the one-line drawing.

a. 69KV feeder line

b. 1200AMP–69KV disconnect switch

c. 69KV arresters (Show leading away on a horizontal line.)

d. 69KV oil circuit breaker

e. 69KV/14.4KV 21/28-MVA transformer, wye-delta

f. 15KV 600AMP air circuit breaker with male and female connectors

g. 5000AMP 15KV disconnect switch

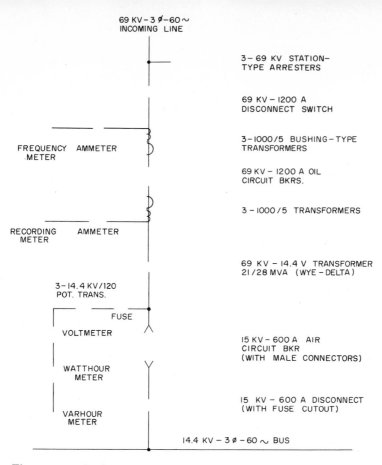

69 KV – 3 ∅ – 60 ∼
INCOMING LINE

3 – 69 KV STATION-
TYPE ARRESTERS

69 KV – 1200 A
DISCONNECT SWITCH

FREQUENCY   AMMETER
METER

3 – 1000/5 BUSHING – TYPE
TRANSFORMERS

69 KV – 1200 A OIL
CIRCUIT BKRS.

3 – 1000/5 TRANSFORMERS

RECORDING   AMMETER
METER

69 KV – 14.4 V TRANSFORMER
21/28 MVA (WYE – DELTA)

3 – 14.4 KV/120
POT. TRANS.

FUSE

VOLTMETER

15 KV – 600 A AIR
CIRCUIT BKR
(WITH MALE CONNECTORS)

WATTHOUR
METER

15 KV – 600 A DISCONNECT
(WITH FUSE CUTOUT)

VARHOUR
METER

14.4 KV – 3 ∅ – 60 ∼ BUS

Fig. 9·12 (Prob. 9·3) Incomplete one-line diagram of substation.

Fig. 9·13 (Prob. 9·4) Sketches of substation sections. (a) Section AA (incomplete) through substation yard. (b) Section BB (incomplete) through substation yard.

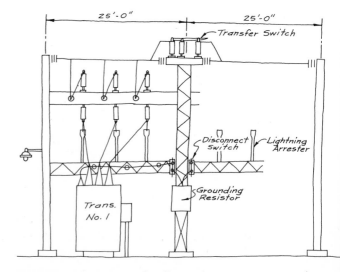

25'-0"   25'-0"

Transfer Switch

Disconnect
Switch

Lightning
Arrester

Grounding
Resistor

Trans.
No. 1

(a)

Section A-A

h. 1000/5 current transformers (Show leading horizontally to ammeter and watthour meter.)

i. 22,000KW turbogenerator

j. 15KV/120V potential transformers (Show leading horizontally to voltmeter.)

k. Ground

9·3 Complete the one-line diagram of the substation shown in Fig. 9·12 by adding the missing symbols. Place suitable identification to the right of each symbol, except for meters and fuse.

9·4 Figure 9·13 shows two incomplete sketches of sections through two different substations. The right-hand part in section A-A, which contains transformer No. 2, is identical to the left-hand part shown. In section B-B, the left-hand span is complete, the center span is complete—except for the circuit-breaker arrangement which is identical to that shown at the left—and the right-hand span is identical to the center span. Make a mechanical drawing of each of the sections, completing where necessary.

9·5 Redraw the floodlight detail shown in Fig. 9·14 to a size approximately two or three times that shown in the figure.

9·6 An installation has several motors and a control panel. Draw the following logic diagrams for part of this, as follows:

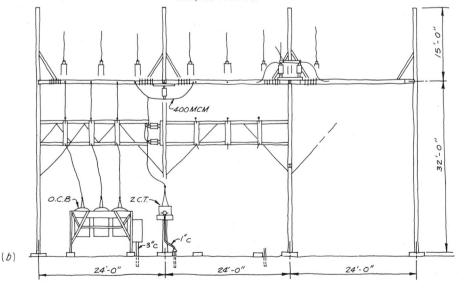

(b)

Section B-B

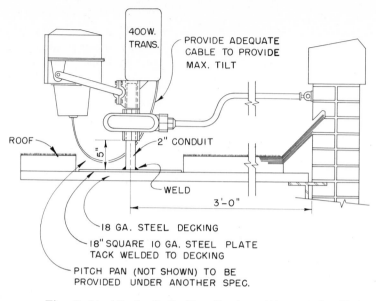

400W.
TRANS.

PROVIDE ADEQUATE
CABLE TO PROVIDE
MAX. TILT

ROOF

2" CONDUIT

5"

WELD

3'-0"

18 GA. STEEL DECKING

18" SQUARE 10 GA. STEEL PLATE
TACK WELDED TO DECKING

PITCH PAN (NOT SHOWN) TO BE
PROVIDED UNDER ANOTHER SPEC.

*Fig. 9·14 (Prob. 9·5) Detail of a 400-watt floodlight.
(Black & Veatch, Consulting Engineers.)*

a. *Control Switch Automatic* and *Safety* must both
be on. *Motor No. 1* will start if the preceding is in
the on-state *or* if the manual control switch is on.
When motor runs, a yellow light goes on at location 2, control panel.

b. *Motor No. 1* will stop if the *Stop Control Switch*
or the *Overload*, or the *Motor Trip* is on. When
the above happens, a green light goes on at location 2, control panel, and a buzzer sounds at the
motor (*MP*-1).

c. For the *Auxiliary Motor* to start, both the *Manual
Control Switch* and the *Safety* must be on. Also
the *Visual* indicator must be on. Use 8½ × 11
paper for each drawing.

9·7 Figure 9·15 shows a gas-burner valve functional
diagram which is a type of diagram used by some
consulting engineers. Make a logic diagram for this
valve and add it to the functional diagram shown.

a. For *fast close* of valve, *Fuel Gas SS valve* must
be open.

b. For *slow close* of valve, *CS* on, *Min Air Flow*, and
Pilot Flame must *all* be in the on-state.

c. For *opening* of *Gas Burner Valve*, the events of
both a and b must all be in the on-state. Note:

264

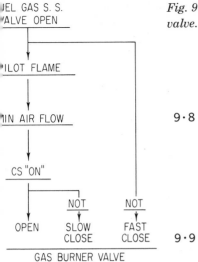

JEL GAS S. S.
VALVE OPEN

PILOT FLAME

MIN AIR FLOW

CS "ON"

OPEN   NOT   NOT
        SLOW   FAST
        CLOSE   CLOSE

GAS BURNER VALVE

Fig. 9·15 (Prob. 9·7) Functional diagram of a gas-burner valve.

Use NOT symbols for closing of the valve. When valve is closed, a green light shows at 33–13. When open, a red light shows at 33–12. Use 8½ × 11 or 11 × 17 paper for each drawing.

9·8 Draw a complete one-line diagram of the electric power distribution in an industrial plant, of which Fig. 9·16 is a diagrammatic sketch. Circuit breakers (oil) are at locations 1 through 14, and 500 HP motors are at *A* and *B*. Possible location of fault currents are at locations *a* through *h*. Use standard symbols throughout. Use 8½ × 11 paper.

9·9 About 60 percent of the one-line diagram for the power distribution of BART is shown in Fig. 9·17. Make a complete one-line drawing of the system with a layout looking like that shown at the upper left. Stations on the Berkeley–Richmond line are (1) RRY, (2) RRI, (3) RCN, (4) RCP, (5) RNB, (6) RBE, and (7) RAS. Stations

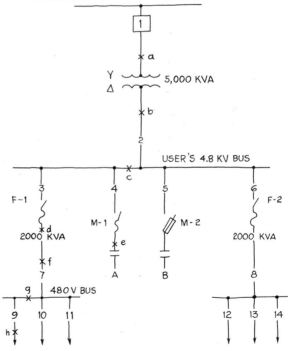

POWER COMPANY SUBSTATION 24 KV BUS

Fig. 9·16 (Prob. 9·8) A diagram of electric power service to a large plant.

on the Mission–Market line are (8) MDC, (9) MBP, (10) MGP, (11) MTF, (12) MSS, (13) MPS, and (14) BTW. Configurations of these two lines are similar to the lines shown. Include all lettering. Use 11 × 17 or 12 × 18 paper.

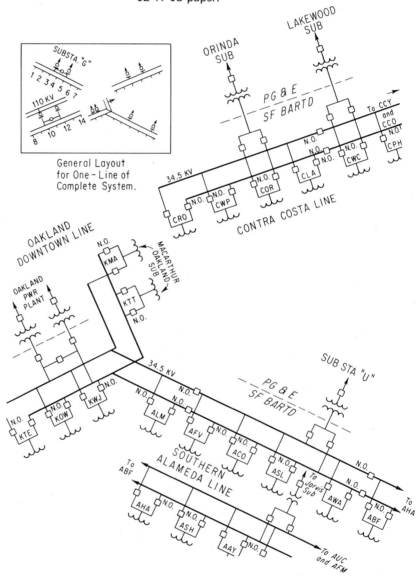

Fig. 9·17 (Prob. 9·9)   Partial one-line diagram of electrification of S.F. Bay Area Rapid Transit System. (Parsons-Brinckerhoff-Tudor-Bechtel, Engrg. Consultants.)

Chapter 10
Electrical drawing
for architectural plans

An integral part of any set of drawings for the construction of a building is the wiring plan or layout. Several standards apply to this type of design and graphical presentation. Symbols for the drawings (other than those used previously in this text) are shown and explained in ANSI Y32.9, "Graphical Electrical Symbols for Architectural Plans," Mil Std 15–3, "Electrical Wiring Symbols for Architectural and Electrical Layout Drawings," and in the *Residential Wiring Handbook* published by the Industry Committee on Interior Wiring Design. The "National Electrical Code," published by the American Standards Association and the National Board of Fire Underwriters, provides design criteria that contain basic minimum provisions considered necessary for safety. Certain political subdivisions (cities and counties, for example) often have their own building codes, which are more restrictive than the NEC.

Persons engaged in the production of electrical drawings for architectural structures must be familiar, therefore, with a number of standards and codes. They should also be conversant with standard terminology and equipment. For the benefit of the reader, we are giving some definitions which will enable him to follow the rest of the chapter with more ease. The following are from Chap. 1, Art. 100, of the "National Electrical Code":

BRANCH CIRCUIT: that portion of a wiring system extending beyond the final overcurrent device protecting the circuit.

BRANCH CIRCUIT, APPLIANCE: a circuit supplying energy to one or more outlets to which appliances are to be connected; such circuits to have no permanently connected lighting fixtures not a part of the appliance.

BRANCH CIRCUIT, GENERAL PURPOSE: one that supplies a number of outlets for lighting and appliances.

CONCEALED: rendered inaccessible by the structure or finish of a building. Wires in concealed raceways are considered concealed, even if they may become accessible by withdrawing.

FEEDER: the circuit conductors between the service equipment (or the generator switchboard of an isolated plant) and the branch circuit overcurrent device.

OUTLET: a point on the wiring system at which current is taken to supply utilization equipment.

PANELBOARD: a panel or group of panel units assembled as a unit, including buses and with or without switches or overcurrent protection, for the control of light, heat, or power circuits of small individual (and aggregate, or total) capacity; designed to be placed in a cabinet or cutout box placed in or against a wall or partition and accessible only from the front.

RACEWAY: any channel designed expressly for, and used solely for, holding wires, cables, or busbars. Includes rigid metal conduit, flexible metal conduit, electrical metallic tubing, under-floor raceways, cellular-floor raceways, surface metal raceways, wireways, and busways.

SERVICE: the conductors and equipment for delivering energy from the electricity supply system to the wiring system of the premises served.

SWITCHBOARD: a large single panel, frame, or assembly of panels, on which are mounted switches, overcurrent, and other protective devices, buses, and usually instruments. A switchboard is not intended to be installed in a cabinet, and it should be accessible from the rear as well as the front.

VOLTAGE (of a circuit): the greatest root-mean-square difference of potential between any two conductors of the circuit concerned. On various systems, such as 3-phase 4-wire and single-phase 3-wire, there may be various circuits of different voltages.

10·1 Simplified and true wiring diagrams

A true wiring diagram shows every wire and its connection in a system, or circuit. Such a diagram is shown in Fig. 10·1a, in which four ceiling light-fixture outlets are depicted, two of which are connected to, and controlled by, individual single-

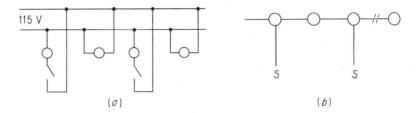

Fig. 10·1 Wiring diagrams of light-fixture outlets on a circuit. (a) True wiring diagram. (b) Simplified, or installation, diagram.

pole single-throw switches. A simplified arrangement of this branch is shown at the right in the same figure. Here, approved symbols have been used for the light outlets, switches, and the wire run, which may be of nonmetallic sheathed cables, armored cables, or any approved method of running conductors between outlets. The two parallel dashes across the wire runs indicate that a two-wire conductor is to be used. Actually, according to the standards, when a two-wire run is to be installed, the dashes may be omitted. If the conductor is to be composed of more than two wires, dashes indicating the number of wires must be provided on the drawing.

10·2 Wiring symbols on a simple floor plan

The architect usually shows the location of lights, convenience and special-purpose outlets, and the desired switching arrangements on a floor plan. For small, simple structures he may draw, or have drawn, the required symbols and wiring arrangements on the same floor plan (Fig. 10·2) that shows all information necessary for the erection of the building. For

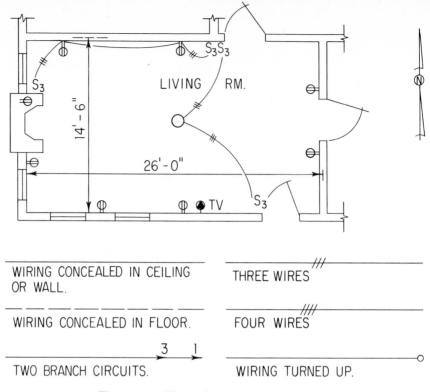

LIVING RM.

14'-6"

26'-0"

TV

WIRING CONCEALED IN CEILING OR WALL.	THREE WIRES
WIRING CONCEALED IN FLOOR.	FOUR WIRES
TWO BRANCH CIRCUITS.	WIRING TURNED UP.

Fig. 10·2   Floor plan of a room with fixtures. outlets, and switches. Standard wiring symbols are below.

larger or more complicated structures he will probably have complete wiring details drawn on separate floor plans, called electrical layouts, or electrical plans. In either case, the simplified type of diagram such as that shown in Fig. 10·1b will be used. This wiring layout will be drawn by an architect, engineer, or draftsman who is familiar with the engineering and building-code requirements.

The living-room plan (Fig. 10·2) shows two circuits: (1) the three-way switching arrangement for the ceiling outlet, and (2) a similar arrangement for the two convenience outlets on the north wall. The outlet symbols, including the special-purpose outlet (indicated for TV antenna), are taken from ANSI Y32.9. Also in accordance with the standard, the wire symbols for the switch-to-ceiling-light runs are drawn with a medium-weight solid line indicating that the wires are to be concealed in the walls or ceiling above. Where no perpendicular dashes are shown across the wires, the conductors must have two wires. With the addition of a little more information about fixtures, the floor plan, of which the living-room plan of Fig. 10·2 is a part, will yield sufficient information for the satisfactory installation of the complete electrical system.

10·3 Separate electrical plans

A plan for the electrical system of a small business building appears in Fig. 10·3. This drawing was one of several, including plans and details for heating, air conditioning, and plumbing, which appeared on a separate sheet of what might be appropriately called mechanical drawings.

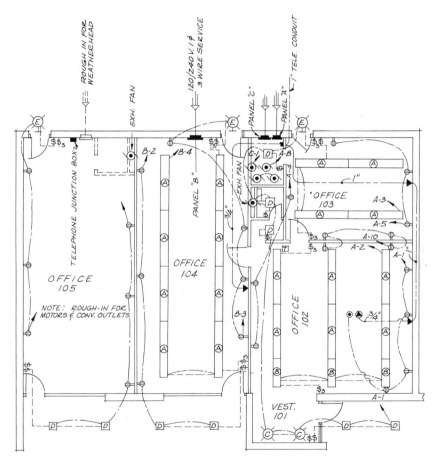

Fig. 10·3 Electrical plan for a small office building.
(Brasher, Spencer & Goyette, Architects-Engineers.)

This electrical plan shows the location of three separate distribution panelboards and the proposed location of a future one. Panel C provides power service—mainly for the motors which run the mechanical equipment; panels A and B supply electricity for lighting and the other electrical needs of offices 102, 103, 104 and the vestibule. Each branch circuit is documented with an arrow pointing in the general direction of the panel and designation such as A-2. This means, for example, that the nine fluorescent light fixtures in office 102 are on the same circuit, No. 2, which is fed at panel A. A separate telephone circuit, enclosed in conduit, is also shown.

A number of symbols that either do not appear in or differ from those shown in ANSI Y32.9 are drawn in this electrical plan. The wall bracket outlets have the four prongs which are still widely used,[1] but which are not shown in ANSI Y32.9. A long and short dashed line is used for circuit legs, regardless of whether the wires are run in the ceiling above, or the floor below, and the short-dash symbol is used for the switch leg. Confusion in the interpretation of these symbols is avoided by preparation of a legend.

10·4 Fixture schedule and legend

Figure 10·4 shows a legend and fixture schedule that accompany the electrical plan of Fig. 10·3. Inclusion of such schedules and legends is the customary practice of architects and consulting engineers who prepare electrical layouts and details for the construction of buildings. The installation of the electrical system is facilitated by the inclusion of letter designation at each fixture symbol and cross-referenced designation in an accompanying schedule. The exact form of schedules has not been standardized. A "remarks" column has been omitted from the original schedule from which Fig. 10·4 was taken in order to conserve space.

A drawing should show intent, and in the most concise manner. Time, money, and argument will be saved if the proper information is placed in the legend. Too often, the man in the field does not see the written specifications. But

---

[1]One explanation for the continued popularity of the four prongs is that many persons feel that the plain circular symbol listed in ANSI Y32.9 may be easily confused with other circular symbols which may appear on drawings.

ELECTRICAL FIXTURE SCHEDULE						
MARK	MFG.	CAT. NO.	MOUNTING	WATTS	LAMP	FINISH
A	FELCO	4039	RECESS	3-95	430 MA	STANDARD
B	FELCO	4033	RECESS	3-38	430 MA	STANDARD
C	LITECRAFT	2305	WALL BRACKET	100	I. F.	SATIN ALUM.
D	PRESCOLITE	488-6600	RECESS	100	I. F.	STANDARD
E	PRESCOLITE	WE-2	WALL BRACKET	100	I. F.	SATIN CHROME

LEGEND :

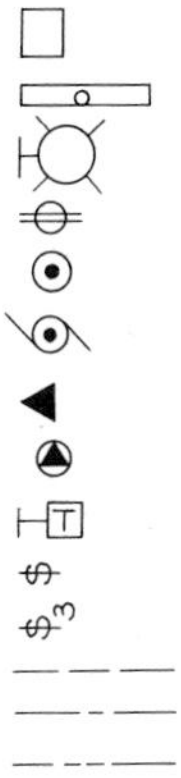

☐   CEILING LIGHT OUTLET - LETTER DENOTES FIXTURE

▭   FLUORESCENT LIGHT OUTLET   "   "   "

⊣○⊢   WALL BRACKET OUTLET   "   "   "

⊖   DUPLEX CONV. OUTLET

⊙   FLOOR CONV. OUTLET

⊙   MOTOR OUTLET

◀   TELEPHONE OUTLET

⬥   FLOOR TELEPHONE OUTLET

⊣T⊢   THERMOSTAT

S   SINGLE POLE SWITCH

S³   THREE WAY SWITCH

— — — SWITCH LEG

— · — CIRCUIT LEG

— ·· — EMPTY CONDUITS

Fig. 10·4 Fixture schedule and legend for electrical plan of small office building. (Brasher, Spencer & Goyette, Architects-Engineers.)

he does see the drawings and the legend. If he sees the symbol for a floor convenience outlet, for example, he knows what the symbol means, but he doesn't know which of the 50 available combinations is required, unless it is specifically stated. Now, if the legend reads "FLOOR CONVENIENCE OUTLET— Frank Adam FB-3," the workman will know exactly what device to install.

10·5 Example of electrical layout

Another electrical layout is shown in Fig. 10·5, and depicts the second floor of a three-story office building. This floor arrangement might be suitable for a drafting or design room. The wiring symbols conform reasonably to the American Standard code. Arrows indicate the circuit and number; half arrows indicate partial circuits.

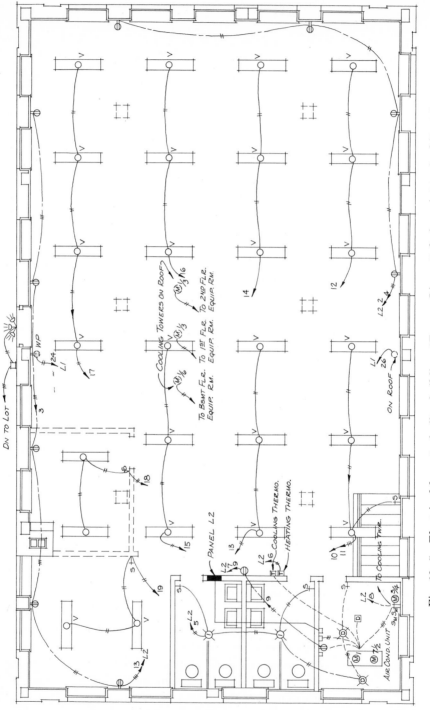

Fig. 10·5 Electrical layout of office building. (Tanner-Linscott & Associates, Architects.)

Included with this electrical plan were a legend, lighting-fixture schedule, panel schedule, disconnect-switch schedule, telephone-circuit riser diagram, and electrical riser diagram. Of these, only the electrical riser diagram is shown in this chapter. Before it is discussed, a word about how energy is distributed throughout a structure seems to be advisable.

As electrical energy is brought into a building, it is usually first passed through a meter. From here it is brought into a main loadcenter. In a small building or residence this loadcenter consists of a fuse box or circuit breaker to which each branch circuit is connected. Through these branch circuits energy is fed to each outlet, lamp, or appliance. In a larger structure the main loadcenter may be a large panelboard, or switchboard, with circuit breakers, disconnect switches, and other controlling devices. From this main loadcenter, electrical energy is fed through large conductors called *feeders* to branch loadcenters, often panelboards, or *panels* as they are sometimes called. From these panels energy is delivered through branch circuits to each outlet, fixture, appliance, or motor. Such factors as location of panelboards, voltage and copper losses, etc., determine the method used in connecting the branch and main loadcenters. Many different interconnecting layouts are used. More than one panelboard may be necessary on the same floor of a large building in order that excessively long runs of branch circuits be avoided. Sometimes separate panels are used for lighting circuits only, and others for motors and machines only. Often, panelboards are used for combinations of types of circuits.

10·6 Electrical riser diagram

A riser diagram shows how electrical energy is distributed through a building from the time it enters the building until it arrives at the branch loadcenters. It is drawn as an elevation and usually is not to scale. An electrical riser diagram for the building whose second-floor electrical plan is shown in Fig. 10·5 appears in Fig. 10·6. This diagram shows the electrical service passing through four meters, then through four disconnect switches. From switches $S1$, $S2$, and $S3$ the current goes to lighting panels $L2$, $L3$, and LB. From the latter two panelboards, energy goes to the lighting and convenience outlets in the ground floor and to mechanical equipment for the ground and first floors. Current travels from panelboard L3 through a subfeeder to L1, which supplies all the electrical service on the first floor, except for air conditioning equipment. Panel L2 supplies electrical energy to most of the circuits on the second floor. However, the electrical layout for this floor (see Fig. 10·5) indicates that outside lighting and mechanical equipment on the roof are connected to panelboards on lower floors.

The panelboard schedule which accompanies the riser diagram includes the following information about each panel:

1 The number of branch circuits to be served
2 Current handling capability (amperage)
3 Name of manufacturer and catalogue number of panel

Similar information is given in the disconnect switch schedule.

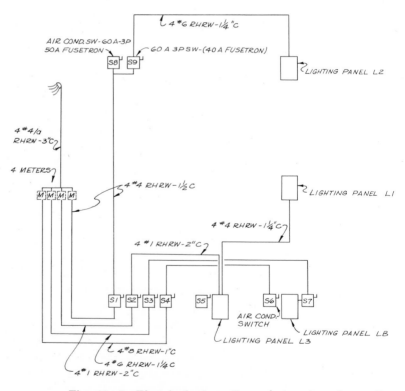

Fig. 10·6 *Electrical riser diagram for three-floor office building. (J. R. DeRigne & Associates, Consulting Engineers.)*

10·7 Electrical drawings for large buildings

The next series of drawings illustrates the type of drawing required for large buildings. The standards are not as much help in the showing of such a complicated system; hence liberal use is made of orthographic projection, pictorial drawing, schematic diagrams, notes, and specifications.[2] A very brief description of the general electrical layout of a large office building will be given by referring to the drawings, which in some instances have been simplified from the original for clarity.

The underground cables of the power and light company enter the building at the upper right, as shown in Fig. 10·7. They immediately are brought into a main service entrance which is shown in some detail in Fig. 10·8. In the main service entrance are current and potential transformers (used for measuring current and potential), fuses, and switches, as shown in the one-line diagram. From the main service entrance the electrical service is fed to two 1,000-kva trans-

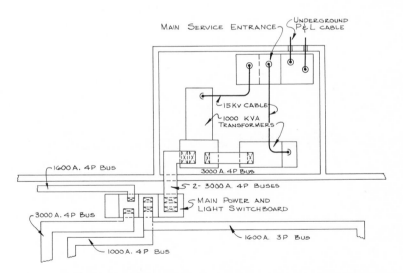

Fig. 10·7 Part of basement plan of large office building showing details of electrical service. (W. L. Cassell, Mechanical Engineer, and Tanner-Linscott & Associates, Architects.)

[2] "The National Electrical Code" (ANSI-C1) contains information that is most helpful in the planning of electrical systems for large buildings.

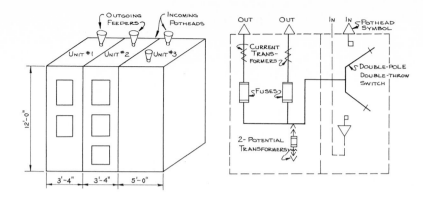

Fig. 10 · 8   Pictorial and one-line diagrams of main service entrance of large office building.

formers (see Fig. 10 · 7) which step the voltage down to a four-phase state that provides 208-volt three-phase service and 120-volt single-phase service. From these transformers current is brought through 3000A 4P (3,000-amp, 4-phase) buses to the main switchboard, shown in pictorial form in Fig. 10 · 9. From this switchgear, electric power is distributed by buses to various parts of the basement, as shown in Fig. 10 · 7. Actually the large copper or aluminum busbars themselves are not shown. Just the bus ducts are drawn—much the same as heating and air conditioning ducts are drawn.

Service to upper floors is fed through the large 3000A 4P bus—which is shown in both Figs. 10 · 7 and 10 · 10 (the riser diagram)—to branch loadcenters on each floor. At each branch loadcenter are a cable tap box, disconnect switches, and panelboards. Here, use is made of metal rectangular raceways (also called "trays") in which connecting cables are placed. From the panelboards service is distributed throughout the floors as explained in the following two paragraphs.

Fig.   10 · 9   Pictorial drawing of main power and light switchboard for large office building.

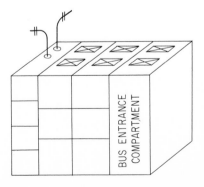

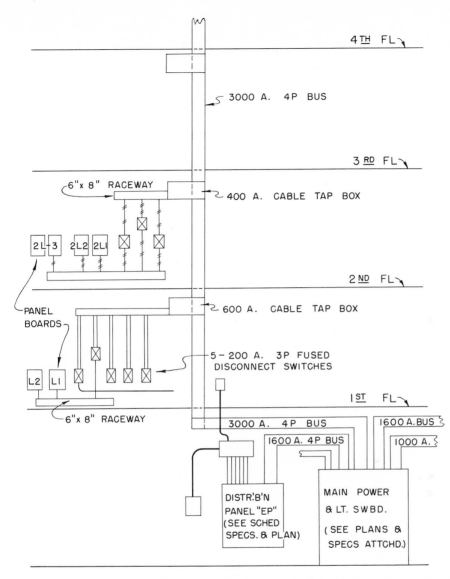

Fig. 10·10 Part of riser diagram for a large office building. (W. L. Cassell, Mechanical Engineer.)

Figure 10·11 shows a part of the electrical plan for the first floor. Shown in somewhat slight detail is the loadcenter with its cable tap box and panels. Of special note is the under-floor duct system shown by means of dotted lines. In these ducts, sometimes called raceways, are placed the wires for

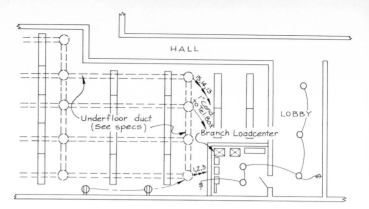

Fig. 10·11 Simplified electrical plan of first floor of a large office building, showing under-floor ducts.

telephone service, 120-volt single-phase electrical service, and 208-volt three-phase service if desired. Outlets may be placed at any point along each of these ducts, and electrical service provided at each of these points. The ducts are covered with 2½ in. of concrete on which additional flooring material is

Fig. 10·12 Under floor duct system being installed. (Walker-Parkersburg.)

often laid. The outlets are usually installed after the latter has been laid, but before the conductor cable has been pushed or drawn through the raceways. However, additional outlets may be emplaced along these ducts after the building has been completed and occupied, although new wiring may have to be poked through the ducts.

Figure 10·12 is a photograph of a raceway system being installed. Note that the raceways are often divided into more than one compartment. It is also possible to have a complete undergrid or cellular floor system. In this type of construction wires can be laid 12 in. apart all the way across the room in both directions. This provides a high degree of flexibility because a telephone, minicomputer, or intercom connection can be placed anywhere in the room.

Details of various electrical hookups and appurtinances are often desirable or necessary. They may take several forms and have different amounts of information. Two detail drawings are shown in Fig. 10·13.

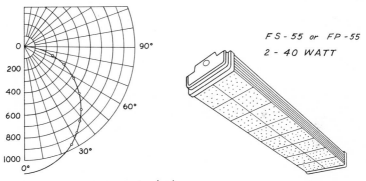

(a)

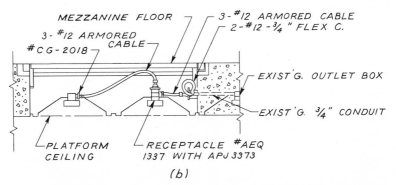

(b)

Fig. 10·13 *Architectural details. (a) Part of luminaire specification. (b) Cabling detail. (S.F. Bay Area Rapid Transit District.)*

10·8 The "National Electrical Code"

The NEC is the overall guiding code which is observed by most architects, engineers, and contractors in the United States. Cities may have their own building codes which are more restrictive than the national code, but the NEC nevertheless is well known to most persons in this field. So that the reader will have an idea as to what some of the provisions are, we are listing some excerpts from the code as follows:

Grounding. A circuit is grounded for the purpose of limiting the voltage upon the circuit. Otherwise, voltages higher than that for which it is designed might occur, through exposure to lightning, for example. The grounding conductor may be connected to the grounded circuit conductor, at any convenient point on the premises, on the supply side of the service disconnecting means. The grounding electrode may be an underground water piping system, the metal frame of a building, or plate, pipe, or rod electrodes.

Overcurrent protection. Equipment must be protected against overcurrent as specified in the NEC. Overcurrent devices shall be located at the point where the conductor to be protected receives its supply, except for certain exceptions. Such devices shall be enclosed in cutout boxes or cabinets, unless mounted on switchboards, panelboards or controllers, or unless part of a specially approved assembly.

Panelboards. All panelboards shall have a rating not less than the minimum feeder capacity required for the load. A lighting and appliance panelboard is one having more than 10 percent of its overcurrent devices rated 30 amp or less, for which neutral connections are provided. Not more than 42 overcurrent devices shall be installed in any one cabinet or cutout box.

Receptacles. Receptacles installed for the attachment of portable cords shall be rated at not less than 15 amp, 125 volts—or 10 amp, 250 volts—and shall be of a type not suitable for use as lampholders. Those located in floors shall be enclosed in floor boxes especially approved for the purpose. Receptacles installed in damp or wet locations shall be of the weatherproof type.

Under-floor raceways. Under-floor raceways may be installed beneath the surface of concrete or other flooring material, or in offices, where they are laid flush with the concrete floor and covered with linoleum or equivalent floor covering. Under-floor raceways shall not be installed where (1) subject to corrosive vapors, or (2) in any hazardous location. The combined cross-sectional area of all conductors shall not exceed 40 percent of the interior raceway, exceptions being armored cable and nonmetallic sheathed cable.

The above are just a very small sampling of the instructions and specifications of the "National Electrical Code," which is periodically revised and brought up to date. A very important section is Art. 90-9, which requires that engineers and others make ample provisions in drawings and specifications for future expansion. This indicates sufficient raceway capacity and space for present and future equipment. The number of wires in a raceway should be limited in order to (1) preclude short-circuiting and breakdown, and (2) allow for future wiring.

10·9 Coordination and organization of drawings

There are usually areas on every drawing which, from the electrical contractor's viewpoint, should be made more comprehensive to help him in preparing his bids. Also the man doing the installing would benefit by having more information about correct or desired procedure, in many situations. However, it is possible to put too much information on the drawing—in trying to answer every question that may arise—and, as a result, confuse everybody from the contractor to the final inspector. Yet when special treatment is indicated, pertinent details should be supplied in order to present a logical and definite manner for the installation.

Many offices which prepare electrical and mechanical drawings for structures have improved their coordination with all crafts doing the work on a job. Such coordination (both before work starts and during the work) avoids conflicts between various mechanical items on the project. Last-minute changes in equipment by manufacturers cause some conflicts and cannot be fully predicted in advance. However, obvious conflicts can be avoided by proper coordination while the drawings are being made.

If a consulting engineer is doing the wiring and mechanical design for an architect, those two offices should also coordinate. Such things as space for raceways, ducts, and air-conditioning equipment must be provided.

10·10 Load computation

Using information supplied in the "National Electrical Code," it is possible to compute the number of branch circuits and feeders required in many types of structures. Let us assume that a residential dwelling has a floor area of 1,490 sq ft, exclusive of basement, attic, and garage, and a 12-kw kitchen range. Using Tables 10·1 to 10·3, we compute the general lighting load and other requirements as follows:

Table 10·1   General lighting loads by occupancies

type of occupancy	unit load per square foot (watts)
Auditoriums	1
Banks	2
Dwellings (other than hotels)	3
Industrial buildings	2
Office buildings	5

Table 10·2   Calculation of feeder loads by occupancies

type of occupancy	portion of lighting load to which demand factor applies (watts)	feeder demand factor (percent)
Dwellings (other than hotels)	First 3,000 or less, at Next 3,001 to 120,000 at Remainder over 120,000 at	100 35 25
Warehouses (storage)	First 12,500 or less at Remainder over 12,500 at	100 50
Others (except hospitals, hotels, and apartment buildings)	Total wattage	100

Table 10·3   Demand loads for household electric ranges, ovens, and counter-mounted cooking units over 1.75 kw rating

number of appliances	maximum demand (not over 12 kw rating)	demand factors	
		less than 3.5 kw rating	3.5 to 8.75 kw rating
1	8 kw (8,000 watts)	80%	80%
2	11	75%	65%
3	14	70%	55%

1 General lighting load

   1,490 sq ft @ 3 watts per sq ft = 4,470 watts

(see Table 10·1)

2 Minimum number of branch circuits required

   a. General lighting load (based on 115 volts): 4,470 ÷ 115 = 39.0 amp. This can be handled by three 15-amp two-wire circuits or two 20-amp two-wire circuits. (NEC recommends one circuit for each 500 sq ft.)

   b. Small-appliance load: (Sec. 220–3 states that two or more 20-amp branch circuits shall be installed to take care of small appliances in the kitchen, laundry, dining room, and breakfast room.)

3 Minimum-size feeders required

computed load	watts
General lighting	4,470
Small appliances (computed at 1,500 watts for each circuit in accordance with sec. 220–3)	3,000
Total (without range)	7,470
3,000 watts @ 100% (Table 10·2)	3,000
7,470 − 3,000 = 4,470 watts @ 35%	1,565
Net computed (without range)	4,565
Range load (see Table 10·3)	8,000
Total net computed	12,565

If 115/230-volt three-wire system feeders are used, the current load will be 12,565 ÷ 230, or 55 amp. Feeder size can then be selected by going to one of several tables in the NEC which give ratings of various types of copper wires, whether they are run singly or in raceways, etc. The feeders in the example might be selected as No. 8 rubber-covered wire of Type R, RU, RUW, or Thermoplastic Type T, wire, or No. 10 asbestos-covered Type AVA, AVL, or others. As mentioned earlier, city building codes may have other requirements not specified in the National Code. Kansas City requires that all wiring be run in conduit for buildings (including houses) that happen to be in the Class A fire district. In other areas, conduits must be used in duplexes and multiple-residency buildings of two stories or more.

SUMMARY   Depicting electrical requirements for residential buildings, offices, and other structures requires a combination of graphical treatment, knowledge of building codes, and written specifications. For small structures, such as houses, it is usually sufficient to show electrical outlets, fixtures, and switches right on the house plans and to indicate what switching arrangements are required. In general, the rest of the circuiting arrangement is left to the constructor. But in larger buildings, it is necessary to depict all the circuit arrangements, including panelboard locations, in order to comply with safety requirements, to allow plenty of flexibility for future equipment and expansion, and to minimize expense of installing electrical work. National Drawing standards ANSI Y32.9 and Mil Std 15–3 (which are identical) are followed by most engineers who prepare electrical drawings. In depicting electrical requirements for buildings, it is necessary or desirable to make wiring plans, riser diagrams, and schedules for fixtures, raceways, conductors, etc. Symbol legends are sometimes used, especially if nonstandard symbols are used in drawings. Pictorial views and orthographic views of certain features and equipment are often quite helpful. Coordination by the person making the drawings with the architect, craftsmen, and equipment manufacturers is highly desirable because it will minimize conflicts and unnecessary installation expenses. Local building codes must be followed by the designer and constructor. When the local code does not cover a situation, the "National Electrical Code" must be consulted. If there is no local electrical code, the NEC should be used.

QUESTIONS   10·1   What authority, or authorities, show the symbols to be used in making a residential floor plan that shows the electrical arrangement?

10·2   Why are not riser diagrams usually drawn to scale?

10·3   Why do we not use bus ducts in residential structures?

10·4   Sketch two different symbols which might be used to show a TV outlet.

10·5   What column headings would you use for a fixture schedule for a small building?

10·6   Why is it not customary to show two dashes across

a circuit line for a two-wire conductor, when three dashes are customarily used to show a three-wire circuit?

10·7   Can you use the same graphical symbol for an electric-range outlet and for a dishwasher outlet?

10·8   Name three different schedules pertaining to the electrical system of a large building that might be found among the drawings for that building.

10·9   Name or describe two situations in which lines for the conductors themselves are not shown on drawings describing the electrical layout of a large building.

10·10   Does ANSI Y32.2 cover the graphical portrayal of the electrical system of a residence?

10·11   Briefly describe three methods of providing under-floor electrical service to an office building whereby electric current can be supplied at evenly spaced intervals throughout or across a room.

10·12   Where does a branch circuit begin? Name two kinds of branch circuits.

10·13   Where are overcurrent devices located?

10·14   What are two reasons for limiting the number of wires in a raceway?

10·15   What are the terminal points of a feeder (or feeder circuit)?

10·16   What is one difference between a switchboard and a panelboard?

10·17   What is the smallest rating a panelboard may have? Or to put it differently, what determines the minimum rating of a panelboard?

10·18   Under what conditions does the NEC say that raceways should not be installed?

10·19   What devices or systems can be used for grounding electrodes? (Name four.)

10·20   What is the minimum rating of a receptacle?

10·21   What can be done to avoid conflicts?

10·22   Why do some architects (or consulting engineers doing electrical design for architects) find it necessary or advisable to add legends to their drawings?

10·23   What sort of information would be contained in a legend for an electrical drawing?

Many of the problems are divided into two parts. Part *a* will be mainly a drawing requirement, and part *b* will involve computational requirements. In some cases, the instructor may require both parts; in other cases he may require only *a* or *b*. Computations do not need to be put on drawing paper. However, they should be neat, orderly, and correct.

10·1   The sketch of the floor plan of Fig. 10·14 shows all electrical outlets and fixtures and the relationships between the fixtures and switches of an apartment.

a. Make a mechanical drawing of this plan using currently approved symbols for wire runs, fixtures, and outlets. Wiring for duplex convenience outlets does not have to be shown except where they are controlled by a switch. Use 11 × 17 or 12 × 18 paper.

b. How many general 20-ampere and appliance 20-ampere branch circuits are required? If the range load is 8,000 watts, and 110/220 three-wire feeders are used, what will be the total current load for the first floor?

10·2   How many 15-ampere branch circuits will be required for the room shown in Fig. 10·2? How many 20-ampere branch circuits would be required for the office layout of Fig. 10·5? (The interior dimensions are 22 ft × 57 ft. Each fluorescent luminaire is 80 watt.)

10·3   Figure 10·15 shows the floor plan of an apartment with all outlets, fixtures, and switches. Switches, for the most part, work overhead lights, but in the living room they operate two of the four outlets.

a. Make a drawing (mechanical or freehand) of this floor plan, adding doors and the necessary wiring. Show TV outlets in the living room and largest bedroom. Use standard symbols. Use 11 × 17 or 12 × 18 paper.

b. A feeder system of 115/230 v three-wire is used and the kitchen range is rated at 11 kw. How many 15-amp branch circuits will be required

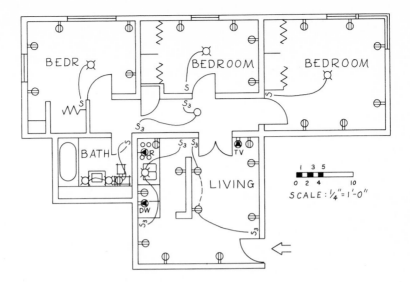

Fig. 10·14 (Prob. 10·1)   *Plan of an apartment unit.*

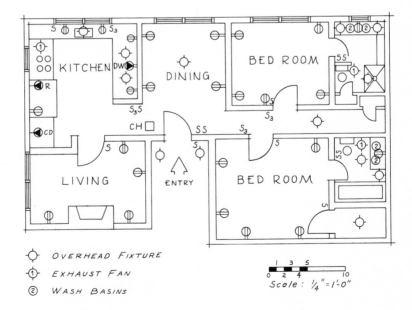

Fig. 10·15 (Prob. 10·3)   *Floor plan of luxury apartment.*

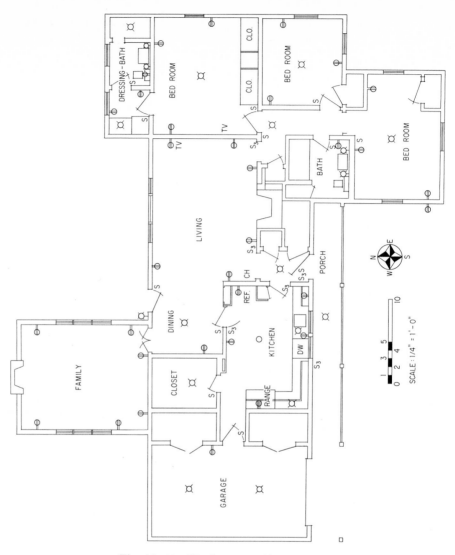

Fig. 10·16   (Prob. 10·4) Floor plan of three-bedroom residence.

for the apartment? What will be the total current load?

10·4   Figure 10·16 is the floor plan of a one-story residence.

a. Draw this plan to a scale of ¼″ = 1′–0″. Show all switching arrangements as you think they ought to be. At least two duplex outlets in the living room are to be controlled by switches at the north entry. Use standard symbols throughout the drawing.

b. The kitchen range has an 11KW rating and 220-volt three-wire feeders are used. How many branch circuits (general and appliance) will be required? (What will the difference be if central air conditioning is installed?) What will be the total current load to be used for computing the size of the feeders?

10·5 Figure 10·17 is the first floor plan of a two-story house with all outlets and fixtures shown.

a. Draw this plan to a scale of ¼″ = 1′-0″, and show all switching arrangements as you believe they should be made. Use correct symbols as

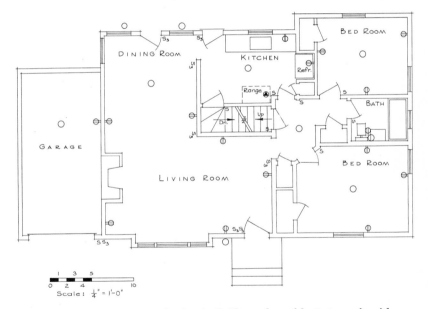
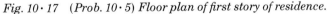

Fig. 10·17 (*Prob. 10·5*) *Floor plan of first story of residence.*

given in the appendix. Optional: show circuits for the convenience outlets. Use 11 × 17 paper.

b. Assume the house has three bedrooms and a hall upstairs. For an 8KW kitchen range and the other equipment shown, how many and what kind of branch circuits would be required for the house? What is the total energy requirement?

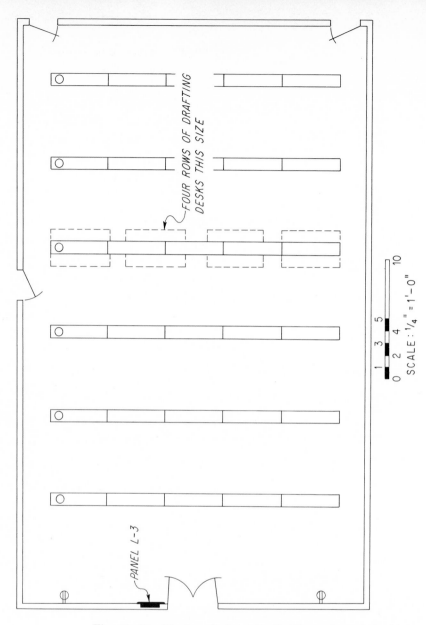

Fig. 10·18 (Probs. 10·6 to 10·8) Floor plan of drafting-design room.

10·6 The lighting fixtures for a drafting-design room are shown in Fig. 10·18. If the fluorescent lights are rated at 80 watts each, estimate the number of branch circuits needed. An electrical outlet should be put near each desk. Four desks are placed under each row of luminaires, making a total of 24 drafting

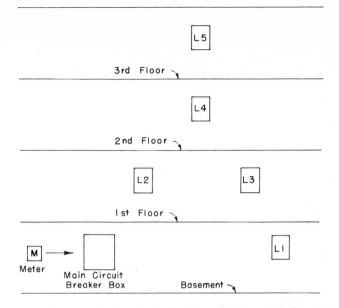

Fig. 10·19 (Prob. 10·9) Section elevation of three-story office building with electrical distribution equipment.

desks. Lights will be turned on and off at panel *L*3. Other circuits go to panel *L*1 on another floor. Make a scale drawing of this room, and show all circuits. Use 11 × 17 or 12 × 18 paper.

10·7 Under each luminaire shown in Fig. 10·18 will be placed a designer's desk. It is desired to have provisions for electric erasers, telephones, and electric calculators at each desk. Make a detailed drawing that includes a system of under-floor raceways which will provide this electrical service to all desks. Assume that telephone conductors and 115-volt conductors can be placed together in a raceway. Use 11 × 17 or 12 × 18 paper.

10·8 Because the telephone company objected to having telephone lines in the same duct with other conductors, it was decided to install a cellular-type, under-floor wiring system for the drafting room of Fig. 10·18. Draw the floor plan to scale, showing a cellular system that will supply 115-volt ac energy to outlets at each desk, and separate cellular ducts for intercom and telephone service to each desk. Overhead lighting may be included in the drawing. Use 11 × 17 or 12 × 18 paper.

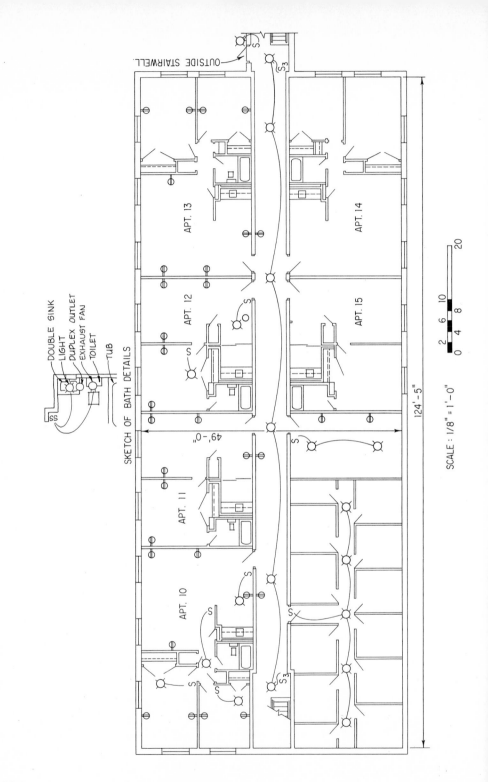

SKETCH OF BATH DETAILS

DOUBLE SINK
LIGHT
DUPLEX OUTLET
EXHAUST FAN
TOILET
TUB

OUTSIDE STAIRWELL

APT. 13

APT. 14

APT. 12

APT. 15

APT. 11

APT. 10

49'-0"

124'-5"

SCALE : 1/8" = 1'-0"

0   4   8
2   6   10   20

10·9   Figure 10·19 shows a section elevation of a three-story building with a basement. Draw a riser diagram for this building as you believe it would be designed. Figure 10·6 will give you an idea of what conductor sizes to use. RHRW (type No.) is a heat-resistant, moisture-resistant wire with rubber insulation. Its current-carrying capabilities in moist locations are as follows:

Fig. 10·20 (Prob. 10·10) Electrical and heating plan of ground floor apartments.

no. of conductor	ampere capacity
1	110
2	95
3	80
4	70
6	55
10	30
12	20
14	15

10·10   The incomplete heating and electrical plan of one floor of an apartment building is shown in Fig. 10·20. Overall dimensions are also shown. Apartments 10, 13, and 14 are identical except that they are reversed. The same is true for apartments 11, 12, and 15.

a. Make a complete scale drawing of this plan, showing all circuiting and switching arrangements for all the rooms and the hall. Appliance loads are 8 KW for ranges, 2 KW for dishwashers, and 4.2 KW for washer-dryers. These appliances will be located in the kitchen of each apartment. Their exact locations and outlets are not shown. (Use 12 × 18 paper for a ⅛″ to 1′–0″ scale.)

b. If a copy of the NEC is available, compute the minimum number of branch circuits and the minimum size sub-feeder for each apartment. Compute, also, the main feeder for this floor and the rating of the panel, if one panel is used.

10·11   Letter a fixture schedule to accompany an electrical plan. Such a fixture schedule might include the information shown in Prob. 1·7, Chap. 1.

Chapter 11
Graphical representation
of data

The graph is one of the most effective tools of communication that any technically trained person can wield. With a graph, he can bring order to a collection of data and present it in a picture form that tells a story. Also with a graph, he can compare the performance of two or more related items or processes, and he may be able to make certain computations not practicable by algebra, analytic geometry, or calculus. In development and research the graph is used to determine the relationships of two or more variables, to compare laboratory data with theory, and to determine if test data are accurate and reliable.

Because this information must be presented honestly, accurately, and as clearly as possible, skill and judgment are required to make a good graph. Therefore, an engineer or draftsman should develop as much skill and knowledge within this area as he would in any other area or type of technical drawing. The fundamental principles of graphical representation are:

1   Graphs should be truthful representations of the facts.
2   Graphs should be clear, easily read and understood.
3   Graphs should be so designed and constructed as to attract and hold the attention.

Figure 11·1 is a good example of a well-drawn technical or engineering-type of graph.

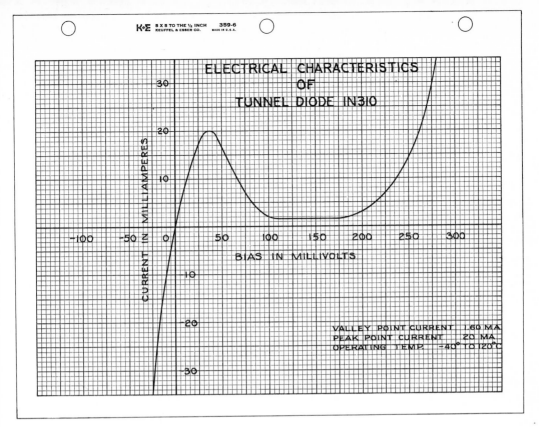

Fig. 11·1   A typical graph plotted on rectangular coordinate paper.

11·1 General concepts in preparing graphs

One of the most difficult problems in constructing a graph is that of choosing the horizontal and vertical scales. As an example of this problem, Fig. 11·2 is presented. In this figure the same "curve" is shown on four different graphs, each having different arrangements of scales; thus, it may be difficult to recognize that the same data are being presented. Graph a is the best of the four for two reasons: (1) the data are presented more honestly, and (2) better use is made of the available space than in any of the other three drawings. This problem of the selection of scales can often be solved by using commercially prepared graph paper and making wise use of space available. However, it may not always be possible or desirable to use such paper.

Another problem in graph construction is that of deciding what type of graph to make. In other words, the problem might be, "What type of graph paper should be used?" There are many different types of graphs in use today, and many different kinds of graph paper. Most graph papers are printed on 8½ × 11-in. paper, although larger sizes are available. Lines come in black, blue, green, orange, and red for many graph styles. Grids are available in rectangular coordinates, and polar and probability coordinates, to name several examples.

Most technical and engineering graphs are drawn on *rectangular coordinate* paper. (It is also called *arithmetic, rectilinear, cross-ruled,* and *square-grid* paper.) Typical spacings for this type of paper are 5, 10, and 20 lines (or spaces) to the inch and 10 to the centimeter. Other spacings such as 4, 6, and 8 lines to the inch are available, but are not in wide usage. The graph of Fig. 11·1 was made on paper that has 10 lines to the inch, as are several other examples in this chapter. If it is desired to make blueprints, Ozalid prints, or other similar types of reproductions, thin, translucent graph paper can be used. If the graph paper has blue grid lines, these will not appear in the reproduction, but orange, red, or black lines will show up on a print. The graphs in this chapter

Fig. 11·2 Graphs having different scales on which identical data have been plotted.

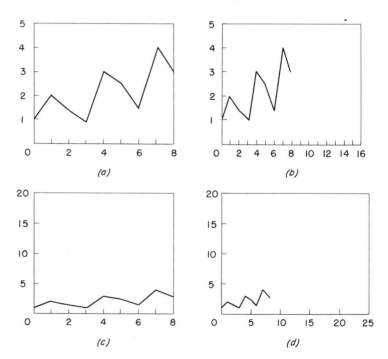

(a)   (b)

(c)   (d)

which have the grid lines showing have been drawn on red-line graph paper.

In order to make good reproductions (by Ozalid, Thermo-fax, xerography, and other reproduction processes), lines and lettering must be made heavy and dark in order to "stand out" from the grid lines. Ink drawings give the best results, but dark, heavy pencil work will provide legible copies in some reproduction processes. In order to make lettering stand out even more, one can do the lettering on heavy, white paper, then paste the paper on the graph. This has been done with the title on Fig. 11 · 5 and in several other graphs in the chapter. This is effective, but not necessary, and will not reproduce on blueprints or Ozalid prints. Sometimes the thin grid lines on a graph can be used as guidelines. This has been done with the supporting data (lower right) of Fig. 11 · 1. This lettering should be the smallest lettering on the graph. The $\frac{1}{10}$-in. height between successive guidelines is ideal.

The tunnel-diode curve of this graph falls in two quadrants, making it necessary to arrange the vertical and horizontal scales so that negative as well as positive values can be plotted. By locating the zero point (*origin*) to the left of center, it was possible to draw the desired portion of the curve, show the scales and their captions, and provide space in the upper part for the title and in the lower right for the auxiliary, or supporting data. The x and y scale values could have been placed around the edges of the graph, but it is believed that they are more appropriately located close to their respective axes, as shown in Fig. 11 · 1.

11 · 2 Selection of variables and curve fitting

Data to be plotted graphically are generally available in tabular form, with each point having two coordinates as follows:

point	coordinates	
1	0	0
2	10	2.6
3	20	3.8
4	30	5.3
5	40	7.8
6	50	10.1

One set of coordinate values must be plotted on the horizontal, or X axis, (or *abscissa* as it is often called) and the other set of coordinates must be plotted along the vertical, or Y axis, (or *ordinate*). Standard practice is to plot the *independent* variable *horizontally* and the *dependent* variable *vertically*. The independent variable is that variable which the operator can control during a test, if one can be controlled. In some cases where a variable cannot be controlled, one variable is arbitrarily selected. *Time,* for instance, is generally considered to be the independent variable. A glance at the coordinates, appearing above, shows that the first set of coordinates progresses at even intervals of 10. It is obvious that the operator or observer was able to take readings at 10, 20, etc, either by controlling the variable or, if it were a natural phenomenon such as time, by taking readings at intervals convenient to him. Those coordinate values in the *left* column, then, represent the independent variable.

After the graph is laid out and the points are plotted, the problem of drawing, or "fitting," the final curve presents itself. Whether to draw a smooth curve or straight lines between successive points depends on several factors. Some of them are as follows:

1 Most physical phenomena are "continuous." This means that the curve showing the relationship between such variables should be smooth—with few inflections—and should pass through or near plotted points.
2 Data backed up by theory should be represented by a smooth curve. (However, if plotted points are not abundant, straight lines are often drawn between points, as with instrument calibration.)
3 Discrete, or discontinuous, data—representing a discontinuous variable having discrete increments—should be shown by joining successive points with straight lines.
4 Observed data not backed by theory or mathematical law should be represented by point-to-point straight lines, unless continuity can be definitely established.

Figure 11·1 includes a curve representing continuous data. Figure 11·2 has a curve which follows the discontinuous relationship of periodic observations. Figure 11·3a shows a theoretical curve and points taken from actual field data. Such

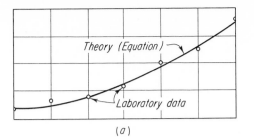

(a)

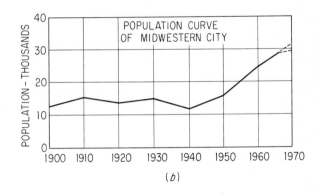

POPULATION CURVE
OF MIDWESTERN CITY

(b)

Fig. 11·3 Examples of curve fitting. (a) A situation when a curve should be drawn through or near points plotted from laboratory data. (b) Census figures, taken once every ten years, yield discontinuous, or discrete, data. Straight lines should be drawn from point to point.

points should be joined by a smooth curve. Figure 11·3b is a typical example of discontinuous data.

11·3 Curve identification

Not infrequently, it is necessary to put more than one curve on a graph. The curves must be drawn or labeled so that they can be easily identified and distinguished from one another. There are three methods by which this can be done:

1 Clearly label each curve
2 Use a different type of line for each curve
3 Use different plotting symbols for each curve

Often two or more of these methods are combined. For example, in Fig. 11·4 different lines and plotting symbols have been used, except for the *composite* curve, which is a sort of weighted average of the other curves. A similar identification system has been used in Fig. 11·5. The difference between the two figures is the method of labeling curves. In Fig. 11·4, a *legend* in the upper right corner identifies each line. In Fig. 11·5, each line is identified by means of a title, or caption, and a leader pointing to the curve itself. Both methods are widely used.

11·4 Zero point location

In a great majority of cases, line graphs are drawn with the *origin* at the lower left corner. Figures 11·2 and 11·4 are in this category. In other cases, the vertical scale begins at zero, but the horizontal scale does not. Figure 11·3*b* is a good example. Most engineers feel that the *Y*-scale should begin at zero; or conversely stated, they feel that starting the *Y*-scale with a figure *not* equal to zero tends to distort the picture.

Sometimes, however, the values to be plotted on the verti-

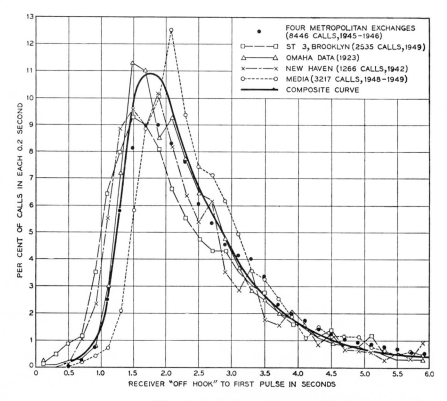

Fig. 11·4 A composite curve. The heavy curve is a composite of the other four curves. (Bell Telephone Laboratories.)

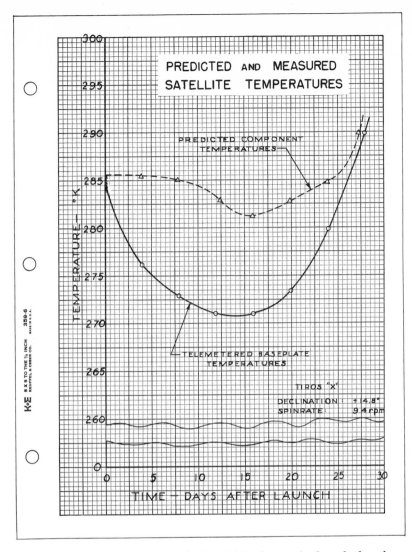

Fig. 11·5 A graph in which the vertical scale has been "broken" to permit greater vertical excursion.

cal scale are such that starting the scale at zero will produce a plotted curve that is too flat and inaccurate for working purposes. Figure 11·5 is an example of this problem. Note that the temperature plots of the weather satellite fall between 270 and 290 degrees, Kelvin. By "breaking" the vertical scale between zero and 260 degrees, we were able to establish ordinate values which provided sufficient vertical latitude ("excursion") for accuracy and comparison of the two curve

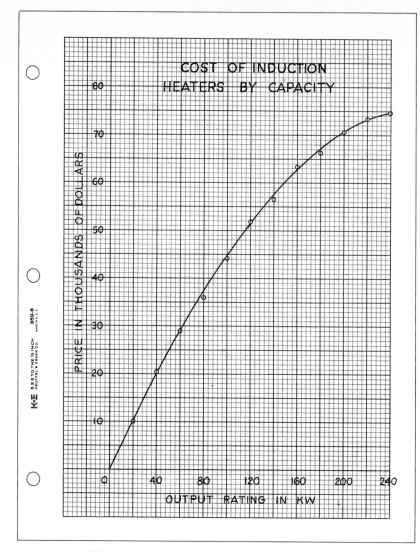

Fig. 11·6 Another typical engineering graph.

shapes. There are other techniques for accomplishing the same result, the point being to warn the reader that the vertical scale is not all there or that it starts at something other than zero.

11·5 Steps in construction of an engineering graph

The following steps illustrate how a graph, as drawn in many engineering offices, is constructed. Such a graph is shown in Fig. 11·6.

1 Arrange the data in table form, preferably in a logical order, from the smallest figure to the largest. This will make plotting easier and faster, and will clearly show the upper and lower limits for each variable.

2 Determine which variable will be the independent variable and which the dependent, for the very practical reason that proper coordinate paper must be selected and a decision made whether the long dimension of the paper will be up and down or sideways on the drawing board. These will probably depend on which variable is which.

3 Select the graph paper that will best show the curve and that will accommodate the points to be plotted taking into account their extreme values. This may include a trial plotting of values along the horizontal and vertical axes. In the case of Fig. 11·6, we selected rectangular coordinate paper having fine lines spaced $\frac{1}{10}$ in. apart, and heavier lines $\frac{1}{2}$ and 1 in. apart.

4 Locate the zero point of each scale (origin of graph), allowing for margins within the grid portion of the paper at the bottom and left side. One-inch margins are common because they allow room for binding the graph at the left side, and sufficient room for scale values and descriptions of these scale values.

5 Letter in the values along and outside the two "baselines" provided in step 4. Standard practice is to use multiples of 1, 2, 5, or 10. In Fig. 11·6, multiples of 10 (each inch representing $10,000) were used on the vertical scale, and multiples of 2 (each half-inch representing 20 kw) were used on the vertical scale. Do not put these numbers too close together.

6 Letter in the descriptions of the scales close to the figures, as shown in Figs. 11·6 and 11·7. Vertical lettering should be used, and each description should be centered between the zero point and the figure at the other end of the baseline. Each description should tell what the scale values show, and what units the scale values represent. Typical standard abbreviations are: kw, amp, cps or Hz (cycles per second), and kHz (kilocycles per second).

7 Plot the points with a sharp pencil or pricking instrument. It is a good idea to circle these points immediately after plotting so that they will be easily seen when drawing the

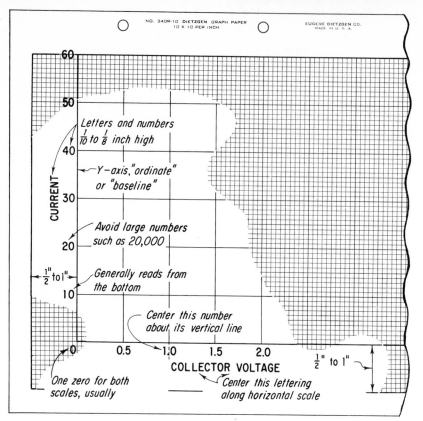

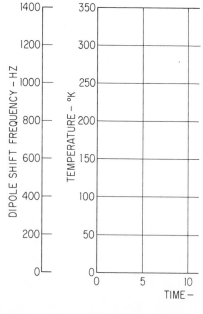

Fig. 11·7 (a) Layout details for the construction of a graph. (b) When two or more scales are required.

curve later. If the circles are to be shown permanently, as is so often done with experimental data, they should be hollow and from $\frac{1}{16}$ to $\frac{1}{10}$ in. in diameter.

8 Draw in the curve. If the curve is not of the straight-line variety, many persons prefer to sketch it lightly freehand until the desired result is obtained, then to put it in with a heavy pencil line, using an irregular curve. If plotting symbols are used, the curve should not be drawn through the symbols. If a symbol is in the path of the curve, the latter should come up to and just touch each side of the symbol.

9 Place the title on the graph. Titles—which are clear, yet as brief as possible—are commonly placed either above or below. This lettering should be vertical uppercase and larger than the numbers and letters describing the vertical and horizontal scales. Some firms require that the fine, preprinted grid lines be erased from around titles and borders drawn around the lettering.

10 Place supporting data, if desired or required, on the graph. Such data may include date of preparation, site of tests or observations, name of person or party making the tests or graph, source of data if not original, equations, and simple circuit diagrams. Such data are often placed in the lower right-hand area of the graph, but are sometimes placed elsewhere if circumstances require. (See Figs. 11·1 and 11·5.)

11 Complete the graph. This may include inking curve, borders, and lettering if inking is required. Sometimes letters and figures can be typed on graph paper with standard typewriters or varitypers if special ribbon is used.

A graph should be made interesting and clear through the use of different line weights. The curve should be the heaviest line on the graph, the border(s) or baselines the next heaviest, and grid lines (if not preprinted) the lightest.

11·6 Drawing a smooth curve

Drawing a curve generally involves two steps:

1 Sketching, freehand, a trial curve through (or near, as the case may be) the plotted points

2 Drawing the finished curve along the trial curve, using pencil or pen and a plastic curve or spline

Some of the plastic (often called *irregular* or *French*) curves and their usage are shown in Fig. 11·8. In 99 cases out of 100, it will not be possible to select a plastic curve

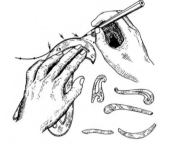

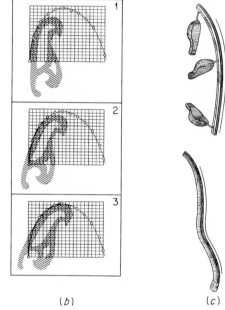

Fig. 11·8 Irregular curves, splines, and use of the curve. (Parts (a) and (c) from Frank Zozzora, "Engineering Drawing," 2d ed., Mc-Graw-Hill Book Company, New York, 1958. Part (b) from Thomas French and Carl Svenson, "Mechanical Drawing," 6th ed., Mc-Graw-Hill Book Company, New York, 1957.

(*a*)   (*b*)   (*c*)

that will match the plotted curve for its entire length. The next-best solution is to try a curve (if more than one is available) or that part of a curve (if only one is available) that will fit as much of the trial curve as is possible. (If the curve is to be inked later, it is a good idea to remember which parts of the curve were used at different locations along the final curve shape.)

Curve drawing is usually more satisfactory if the pencil is used on the edge of the curve that is away from the person who is drawing. This is somewhat analogous to using the upper (far) edge of a T square.

11·7 Scales and their titles or captions

There are good and bad ways to organize the numbers and descriptions along the baselines of a graph. Here are a few examples:

0	10	20	30	40	50
0	0.2	0.4	0.6	0.8	1.0
0	50	100	150	200	250

poor selection of numbers

0	10000	20000	30000	40000	50000
0	30	60	90	120	150
0	7	14	21	28	35

good title or caption organization

RESISTANCE IN OHMS
RESISTANCE IN THOUSANDS OF OHMS
RESISTANCE — THOUSANDS OF OHMS
RESISTANCE IN MEGOHMS

satisfactory but sometimes confusing

RESISTANCE—OHMS $\times$ 10^3
PARTS PER FT^3 $\times$ 10^6
TRIGGER CHARGE—10^{-10} COULOMBS
(The numbers shown are all positive.)

11·8 Families of curves

Figure 11·9 shows a *collector-characteristic* curve of a transistor which was obtained by varying the voltage and measuring collector current for several values of base current. This family of curves can be used for the determination of transistor performance in a common-emitter circuit and for the calculation of other useful parameters. For example, it is often desirable to know the ratio of the d-c collector current, I_c, to the d-c base input current, I_B. This *current gain* is typically around 49 or 50, calculated as follows:

$$\beta = \frac{I_c}{I_B} = \frac{0.981}{0.021} = 49$$

Other useful characteristics which can be obtained from such a table and other information supplied by a manufacturer are frequency cutoff, breakdown voltage, reach-through voltage, and storage time. When five or more curves are

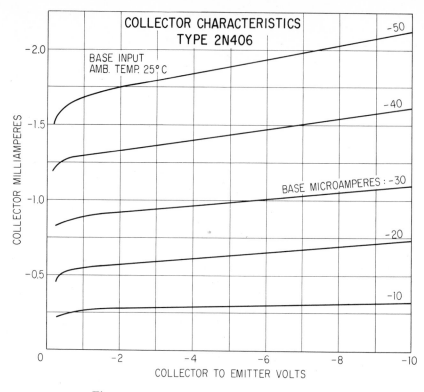

Fig. 11·9   *A family of curves. (RCA.)*

placed on a single graph, it is not very practical to draw a different type of line for each curve. Proper identification of each line is usually accomplished in the manner shown in Fig. 11·9.

11·9 Graphs for publication

When graphs are to appear in printed matter, such as books or technical journals, they are usually not drawn on commercial graph paper. ANSI Y15.1, "Illustrations for Publication and Projection," is an appropriate guide for such cases. Covered are such points as minimum letter size and line weights before reduction to legible and clear size. A standard outline proportion of 6¾ × 9 in. (¾ : 1) is recommended. No more coordinate lines should be drawn than are necessary to guide the eye. Because additional discussion is presupposed, supporting data and equations should be omitted. In the majority of published graphs, plotting symbols are not shown, although some do include these symbols. Figures 11·9, 11·10, and 11·14 are good examples of this type of graphical presentation.

In short, the published graph is a rather simple construction, uncluttered by minute data, yet very clear and legible. In order to have clarity and legibility, it might not include all data necessary to tell the whole story.

11·10 Line graphs on other types of graph paper

As mentioned previously in this chapter, other kinds of commercially printed graph paper are available, and used by many engineers and scientists for various reasons. Figure 11·10 exhibits a response curve for an amplifier over a range of frequencies going from 100 Hz to 5,000,000 Hz. The

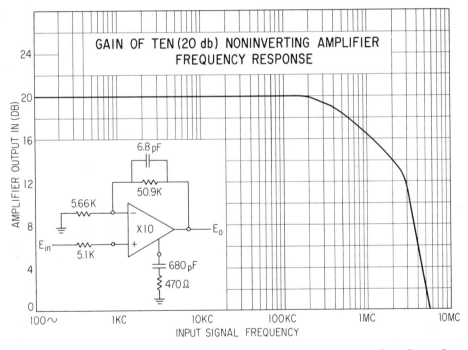

Fig. 11·10 A frequency-response curve plotted on five-cycle semilogarithmic paper. The range on the horizontal scale is from 100 cycles per second to 10 megacycles (10 million cycles per second). The new term for cycles per second is hertz, abbreviated Hz.

vertical scale (not logarithmic) is measured in decibels, or units of volume. This response curve, as are most amplifier-response curves, is plotted on five-cycle semilogarithmic paper, in order to plot accurately along the great ranges of frequencies. It would be impossible to plot numbers over such a large range on linear (rectangular coordinate) paper. This is one reason for using paper with one or more logarithmic scales.

Another reason for using different kinds of graph paper is to achieve a pattern that yields a "straight-line curve." Sometimes, points that yield a curved pattern on one kind of paper yield a straight-line pattern on another kind of paper. The following will result if a straight-line curve can be drawn:

1 Future prediction (extrapolation) is easier if a straight line is used.
2 Two curves can be better compared if they are straight-line curves than if they are otherwise shaped.
3 The slope and equation of the line (and therefore of the data which produced the line) can be obtained graphically.

If the logarithmic scale is oriented vertically and the linear scale horizontally, a curve representing a constant *rate* of change will plot as a straight line. Such a curve would result if we were to plot the 2.8 percent annual increase in the rate of change of our output per man-hour since World War II.

11·11 The logarithmic scale

The logarithmic scale is a functional scale in which the distances are laid out to equal the function (logarithm), but the numbers at these distances are those of the variable. A logarithmic scale from 1 to 10 is called a *cycle*. The left edge of the scale is marked 1 (the logarithm of 1 is zero), and the right edge 10 (the logarithm of 10 is 1), and we have a *unit* scale from the log of 1 (zero) to the log of 10 (1). This unit scale is called a cycle, but we can scale intermediate distances like 1.5, 2, 3, 5, etc., and show those numbers at those points. A logarithmic scale from 10 to 100 or 100 to 1,000, etc., is also called a cycle, or *modulus*.

It is impossible to have a zero showing on a logarithmic scale. If the scale includes the number 1 at either end, the zero is there *graphically* because the logarithm of 1 is zero. Also, the logarithm of zero is minus infinity. This would be impossible to plot graphically.

It is also worth noting that *interpolation* between marks on a logarithmic scale must be done *logarithmically*. When using logarithmic graph paper one does not have to be concerned about this.

11 · 12 Equations of straight-line plots

The following equations will apply in the situations indicated:

equation	type of coordinates
$y = mx + b$	Rectangular
$y = \dfrac{x}{mx + b}$	Rectangular $\left(\dfrac{x}{y} \text{ values plotted against } x \text{ values}\right)$
$y = bx^{m}$	Logarithmic
$y = b\,(10)^{mx}$ or $y = b\,(e)^{mx}$	Semilogarithmic (logarithmic scale vertical)
$y = m \log x + b$	Semilogarithmic (logarithmic scale horizontal)

In the above equations, m represents the *slope,* and b the *intercept.* We will show how to get these quantities in the next two examples.

11 · 13 Use of logarithmic paper

Logarithmic paper (both scales are arranged logarithmically) is used for reasons similar to semilogarithmic paper. If the range of plotted values is large in both the x and y directions, logarithmic paper, with the proper number of cycles, can be used very much as semilogarithmic paper was used to accommodate the large range of frequency values used in Fig. 11 · 10. As in the case of semilogarithmic graph paper, logarithmic paper is available, having anywhere from one to five logarithmic cycles, and anything in between. The most-used papers have the same number of cycles in the horizontal and vertical directions, but papers are available with different combinations. Usually, however, the length of one horizontal cycle is the same as the length of one vertical cycle.

Another reason for plotting data on logarithmic paper is

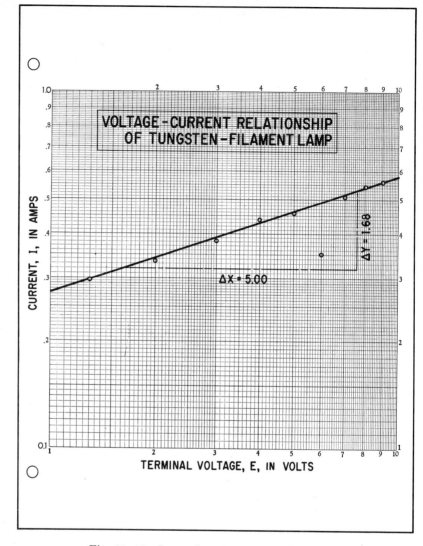

Fig. 11·11 *Data plotted on logarithmic paper. These data yield a straight-line pattern. $\Delta Y/\Delta X$ represents the slope.*

to get a straight-line pattern. It might be found by trial and error that a certain group of data provides a straight-line plot on such paper. Or theory or previous tests with similar data may indicate that a set of plotted points will yield a straight-line pattern. Figure 11·11 contains such a situation. Plotting of points and construction of the graph follow the same techniques, previously explained, for drawing a graph on rectangular coordinate paper. One difference is that the numbers around the edge are usually already printed on the sheet. The draftsman often must add zeros or decimal points to the numbers of the scale.[1] Because these numbers are outside the grid, the scale caption must also be placed outside the grid. Another difference is that the curve itself should not be very thick, if one wants to obtain accuracy in measuring the slope and locating the intercept.

In Fig. 11·11, a right triangle with a 5-in. base has been drawn. Each leg of the triangle has been measured accurately with a decimal scale to obtain Δx and Δy values. (This is *construction* work, and it is sometimes erased and does not appear on the finished graph.) The important point about measuring the legs of the triangle on logarithmic paper (where the horizontal and vertical cycles are equal in length—which is usually the case) is that a linear scale be used, and that the *same* scale be used to measure each leg. Instead of using a decimal scale, we could have used a 50 scale, 40 scale, quarter scale—or any *linear* scale. Using the logarithmic scale of the graph paper to measure the legs will *not* provide the correct slope.[2] The slope is obtained as follows:

$$m = \frac{\Delta Y}{\Delta X} = \frac{1.68}{5.00} = 0.336 \text{ or } 0.34$$

The *intercept* is found by observing where the line intersects the Y scale where $X = 1$ (remember, the log of 1 is zero) and reading the value along the Y scale. The intercept in Fig. 11·11 is

$$b = 0.272$$

The equation for this line is, therefore,

$$y = 0.272 \, x^{0.34}$$

or, using the abbreviations for the actual units plotted,

$$I = 0.27 \, E^{0.34}$$

[1]Sometimes the printer does not leave enough room for this, or there is not much room for scale titles, presenting a rather awkward situation.

[2]If the horizontal and vertical cycles of the paper are not equal in length, one will have to select points on the line and work with their coordinates to get the slope: $m = (\log Ay - \log By)/(\log Ax - \log Bx)$.

The third decimal-place values have been dropped because there probably is not enough justification for this type of accuracy. Two-decimal accuracy is appropriate, however.

11 · 14 Use of semilogarithmic paper

Figure 11 · 12 illustrates a situation in which a series of plotted points provides a straight-line pattern, as shown by the line.

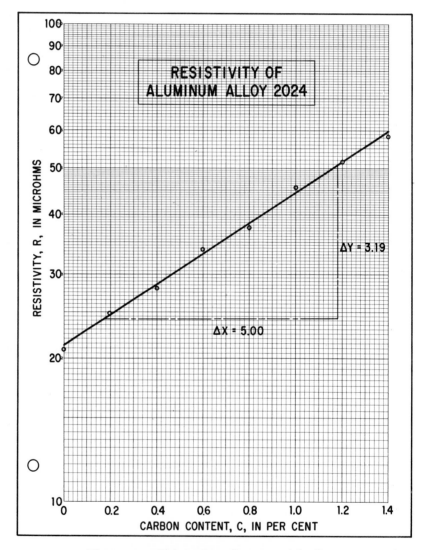

Fig. 11 · 12   Points that give a straight-line pattern when plotted on semilogarithmic paper.

As in the case of the logarithmic graph, previously cited, the scales had to be marked off around the outside edges of the grid because the log scale numbers had already been printed.

The ΔY and ΔX values shown were measured with the same *linear* scale. But because the two scales of a semilogarithmic graph are different, we have to do a little more work to get the correct slope. The additional work (beyond what was done in the case of the logarithmic graph) is to obtain values of *unity* for the X and Y scales. Unity for the X scale is 5.00 in. (the actual distance from $X = 0$ to $X = 1.0$). Unity for the Y scale is 10.00 in. (the distance from log 10 = 1 to log 100 = 2), in this case the length of the logarithmic cycle from 10 to 100. Now, we are ready to determine the slope.

$$m = \frac{\Delta Y/\Delta X}{Y \text{ unity length}/X \text{ unity length}} \qquad \text{measured with the same linear scale}$$

$$m = \frac{3.19/5.00}{10.0/5.0} = 0.319 \text{ or } 0.32$$

The intercept is found by reading the value along the Y scale where the line intersects at $X = 0$:

$$b = 21.3$$

The equation is

$$y = 21.3 \, (10)^{0.32x}$$

or

$$R = 21.3 \, (10)^{0.32C}$$

If it is preferred to use e instead of 10 in the equation, the slope must be divided by 0.434, which is $\log_{10} e$.

$$m = \frac{0.32}{0.434} = 0.74$$

and the equation becomes

$$R = 21.3 \, (e)^{0.74C}$$

If the reader cannot reconcile the graphical method just explained, he can check the slope by selecting two points on the curve and using their coordinates as follows:

$$m = \frac{\log Ay - \log By}{Ax - Bx}$$

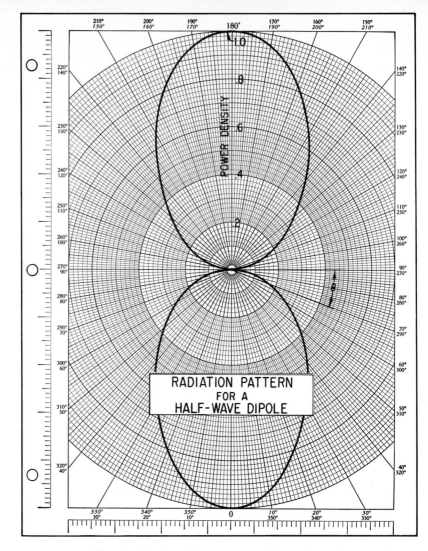

Fig. 11·13 *A radiation pattern plotted on polar coordinate paper. (Zero at bottom.)*

We have done this by selecting points at $A = (1.16, 50)$ and $B = (0.46, 30)$ to get

$$m = \frac{1.699 - 1.476}{1.16 - 0.46} = 0.319 \text{ or } 0.32$$

There are other methods which can be used to obtain, or check, these equations. One method which requires more work but which is more accurate—and which can utilize a digital computer—is the method of least squares.

11·15 Polar coordinates

Because of the directional characteristics of electrical devices
such as lamps, antennas, and speakers, studies must be made
of their performance in different directions. The results of
these studies can be most appropriately shown on polar
charts. These show two variables, one having a linear magnitude plotted on equally spaced concentric circles, and the
other an angular quantity plotted radially with respect to a
pole or origin. The plotted points are usually, but not always,
joined to form a line or curve.

Figure 11·13 shows a polar plot of the power-radiation
pattern of a half-wave dipole, as computed by an analog computer. Notice that the linear scale is shown as units based on
1.0 being the maximum. Notice, also, that the zero is at the
bottom of the graph. The curve has two large enclosures,
called *lobes,* and points of zero magnitude, called *null points,*
at 90° and 270°.

*Fig. 11·14 Another
graph on polar coordi-
nates. (Zero at top.)
(Bell Telephone Labo-
ratories)*

Figure 11·14 depicts the cross-talk coupling between two
antennas at the same location. The two "envelopes" show
the maximum cross-talk obtained for two positions of the
transmitting antenna when the receiving antenna is rotated

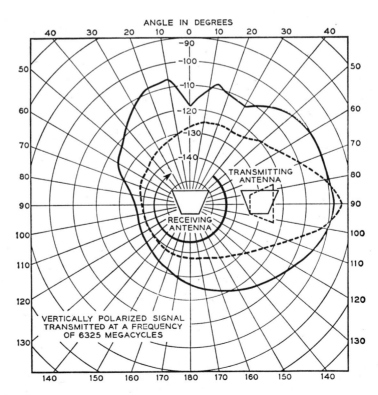

clockwise. Notice that zero is at the top and that degrees increase both clockwise *and* counterclockwise from zero.

In other situations, it is sometimes desirable to plot zero at the right side of the graph and go counterclockwise, much as the mathematician plots values of trigonometric functions. It is possible to buy polar coordinate paper that has the zero at the top or at the bottom. Disk recording devices use polar charting, and instead of radial graduations being in degrees, they are in hours or days.

11·16 Bar charts

Some types of data do not lend themselves well to presentation in line graphs. They might be better suited for display in some other form, such as a bar chart. Also, data in graphical form must often be presented to clients or persons who do not have technical background. Such persons may find bar graphs, pie charts, and pictorial graphs easier to read.

Figure 11·15 is a bar chart in which the bars run up and down, for which reason it may also be called a *column chart*. Its construction is arrived at in much the same manner as a line graph that is plotted on rectangular coordinate paper. General practice is as follows:

1 Several major horizontal grid lines should appear with their scale values.
2 Bars, or columns, should be shaded.
3 Widths between bars should be no wider than the bars themselves and may be less.
4 Bars are often arranged in ascending or descending order, but this is not a requirement.

To make an attractive bar chart, one should consider using preprinted appliqués for shading. Rather than make a scale which would accommodate the United States production of 236,000,000 kw, the author decided to use a scale that would permit the bars of the other countries to be larger and "broke" the United States bar as shown. While not generally done, this is acceptable practice as long as the values are shown at the top of the bars.

Another type of bar chart is shown in Fig. 11·16. This is only one form of a horizontal bar graph. Another form has a zero line running up the middle, with bars representing posi-

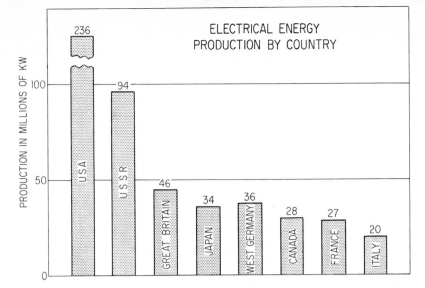

Fig. 11·15   *A bar chart or graph. (Titles for bars are often placed below the bars.)*

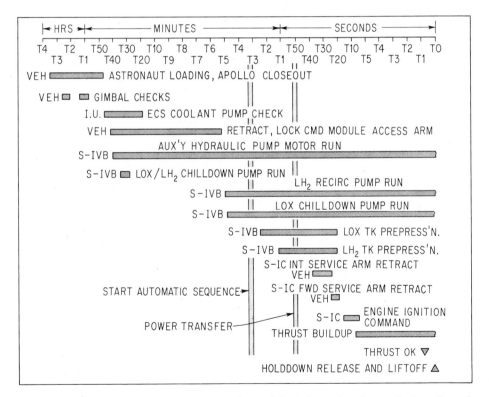

Fig. 11·16   *Horizontal-bar chart showing typical prelaunch events for Saturn–Apollo flight. (NASA.)*

tive values extending to the right and bars with negative values to the left. Bars may also be broken down individually into several parts. For example, if the information were available, we could show what percent of each country's electrical production were (1) hydro, (2) steam generating, and (3) nuclear, on the graph of Fig. 11·15.

11·17 Pie graph

A very popular type of chart, although held in low regard by statisticians, is the pie chart. It is most effective for displaying five to seven items that make up 100 percent. A well-designed pie graph has the largest item in the upper right sector, starting at twelve o'clock, followed by the next largest item, and so on in a sequence of decreasing size. The following principles of good construction apply.

Fig. 11·17 A pie graph. Notice that the material is arranged sequentially by size.

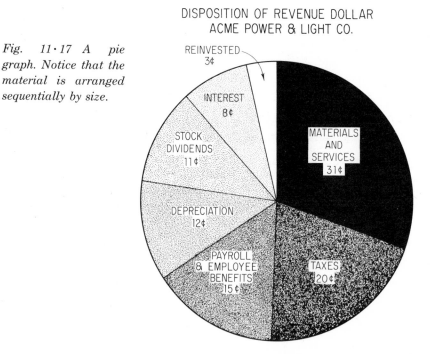

DISPOSITION OF REVENUE DOLLAR
ACME POWER & LIGHT CO.

1. The graph must be large enough to permit lettering within all but the very smallest sectors.
2. Arrangement of items should follow a sequence of decreasing sizes.

3 Items should have a shading sequence of ever-increasing or -decreasing darkness.

4 The largest item should start at twelve o'clock and be on the "clockwise" side.

5 A title should be placed above or below the graph.

6 The percentage, or value, should show in each sector.

11·18 Pictorial graphs

There is some evidence that graphs presented in pictorial form are remembered longer than those drawn in two dimensions. Many different ways of showing graphs in pictorial form have been used, including isometric, dimetric, oblique, and perspective construction. A three-variable pictorial graph has been shown in Fig. 11·18, which has been drawn in isometric projection. Making a graph like this takes quite a bit of time; for

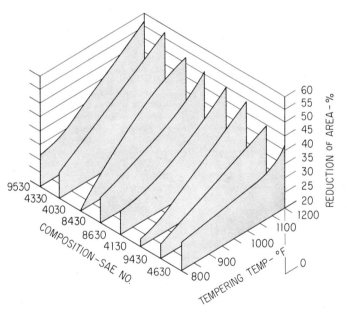

Fig. 11·18 A pictorial graph showing reduction of area of various steels after liquid quenching at various temperatures. (Three variables in isometric projection.)

one thing, lettering is a little tricky. Therefore, one should be sure that the results will justify the extra effort. The material presented in Fig. 11·18 could have been presented in the form of the graph shown in Fig. 11·9, and vice versa.

Another pictorial drawing, shown in Fig. 11·19, is only partly a graph. This shows how ingenuity can be used to combine two different kinds of graphical representation. Other forms of pictorial representation of graphs have been used with success. Bar graphs have often been drawn pictorially. Oblique projection adds depth to the bars and makes possible the addition of more information in the third dimension.

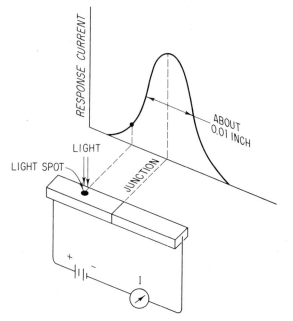

Fig. 11·19 A pictorial drawing (in dimetric projection) showing the response of a PN transistor depending upon where it is illuminated. The curve is a normal frequency-distribution curve.

11·19 Other types of graphic presentation

The graphs shown in this chapter are the type found most often (perhaps 95 percent of the time) in engineering, design, and research offices. Actually, there are a number of other ways to present material graphically, and some of these graphical forms can be used for calculation as well. Some of these graphs have been assigned names as follows:

1 Alignment charts
2 Concurrency charts
3 Conversion scales
4 Network diagrams
5 Holographs
6 Derivative curve ⎫ graphical calculus
7 Integration ⎭

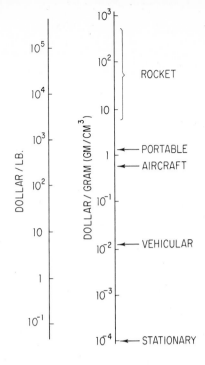

Fig. 11·20 A line diagram which is also a conversion diagram. This shows justifiable cost for removing by miniaturization one pound or gram of unnecessary weight. (From Edward Keonjian, "Microelectronics," McGraw-Hill Book Company, New York, 1963. Used by permission.)

Figure 11·20 is a type of conversion scale that shows the comparative economic advantages of reducing the weights of various electrical or electronics gear. For the other types of graphs, mentioned immediately above, the reader should refer to a good text on engineering graphics, graphics, or graphic science. Some of the better texts have been written by Levins, Luzzader, Mochel, Paré, Rising and Almfeldt, and Vierck.

SUMMARY Graphical construction is a very important phase of technical drawing. A careful survey of the problem is the first and most important step in successful graphic presentation. It should cover a careful consideration of items such as the following:

1 The purpose of the graph
2 The occasion for its use
3 The type of person who is to read it
4 The nature of the data
5 The medium of presentation to be used
6 The reproduction process to be used
7 The time available for preparation
8 Equipment and skill available

A graph should have a sufficiently professional appearance to inspire confidence in the facts presented. Layout and design are as important as quality of drafting. The type of data being portrayed graphically determines whether the curve should be smooth, or a series of straight lines from point to point. Clarity and brevity are the two most important features of any graph. Some graphs can be used for purposes of computation. For example, equations such as $y = mx + b$ and $y = bx^m$ can be obtained if the curves appear as straight lines on the appropriate types of graph paper. Alignment charts, concurrency charts, and other graphical forms are useful in making certain computations.

QUESTIONS

11·1 What determines whether the plotted points on a graph should be connected with straight lines, or be joined by a smooth curve?

11·2 What are two reasons why it is sometimes desirable to have data plotted in one straight line (by using different graph paper) instead of as a curve?

11·3 What determines whether data should be shown by means of a bar (or column) chart, or by a pie chart?

11·4 Name four types of commercial graph paper that are available.

11·5 Along which axis is the dependent variable usually plotted?

11·6 In plotting the data from a laboratory experiment, how would you go about determining which variable is the independent one?

11·7 If it were necessary to draw two curves on one graph, how would you differentiate between these curves?

11·8 Would a graph that has been made into a projector slide show more or less information than a graph for the same data that will be bound in an engineering report?

11·9 Name three purposes for which graphs are, or may be, used.

11·10   What are three different line spacings that may be purchased in commercial rectangular coordinate paper?

11·11   Why is there not a zero point on logarithmic graph paper?

11·12   What do we mean by a "cycle," as it pertains to commercial graph paper?

11·13   List three examples of supporting data that might appear on an engineering graph.

11·14   When using commercial graph paper, where would you place the title? The supporting data?

11·15   If three different weights of lines are to be used in drawing a graph, which parts of the graph will be depicted by which weights of line?

11·16   Name four shapes of plotting symbols that are commonly used in graph work. Which symbol is the most common?

11·17   In what numbers, or multiples thereof, do we usually lay out the baselines of a graph?

11·18   List four abbreviations that are often used to indicate typical units that are placed along a baseline.

11·19   If it is desired to show each point permanently by means of a circular plotting symbol, how large would you make the symbol?

11·20   What characteristic of a heater or antenna can best be shown by means of a polar graph?

11·21   In constructing a pie chart, what two sequences should be observed and followed?

11·22   What would be a reason for using graph paper with blue lines, as opposed to using red-line paper?

11·23   List two sequences of numbers placed along the axis of a rectangular coordinate graph that are considered to be poor sequences.

11·24   Why is it possible to draw a right triangle on a straight-line curve on logarithmic paper and use the linear lengths of the two legs to obtain the correct slope?

11·25   What are six factors that should be considered in the layout of a graph?

PROBLEMS   11·1   The treasurer of Western Utility Power Company has prepared the following information about how the utility spends its "dollar" for presentation at the next stockholder's meeting. Put this information in the form of a pie chart.

Wage and salaries	$0.32
Taxes	0.20
Materials and fuel	0.08
Interest and dividends	0.26
Add. to physical plant	0.14

11·2   Prepare a bar or pie chart for the estimated increase in capacity (in megawatts) estimated by the Public Service Company:

	net addition	*year-end capacity*
1974	$0.98	$ 7.58
1975	0.67	8.25
1976	1.22	9.47
1977	0.77	10.24
1978	0.99	11.23
1979	0.76	11.99

11·3   Plot the hourly demand and normal-capability curves of the interconnected system of Mid-Western Power & Light Co., shown at the top of p. 329.

year	maximum hourly demand (thousands of KW)	normal generating capability (thousands of KW)	year	maximum hourly demand (thousands of KW)	normal generating capability (thousands of KW)
1963	256	263	1968	411	449
1964	265	263	1969	469	437
1965	318	325	1970	501	565
1966	352	364	1971	513	584
1967	381	364	1972	580	742

11·4   The 10-year history of generation data of East States Utilities Company is shown below. Plot the KWH and BTU curves on one graph.

year	KWH generated (billions)	BTU required (billions)	year	KWH generated (billions)	BTU required (billions)
1956	1.32	19.2	1961	1.90	23.0
1957	1.43	20.4	1962	2.14	26.3
1958	1.54	20.4	1963	2.24	26.9
1959	1.71	22.2	1964	2.43	28.6
1960	1.80	23.2	1965	2.62	30.3

11·5   Make a bar chart of the data presented in Prob. 11·3, above.

11·6   Make a bar chart of the data presented in Prob. 11·4 above.

11·7   The following data show the survival probability of nonredundant system and a redundant system with $N = 10^5$ components. Plot both curves and identify each. Failure rate is $f = 10^{-7}$ per hour.

time (hours)	survival probability, P	
	nonredundant system	redundant system
0	1.0	1.0
200	0.2	1.0
500	0.05	1.0
1,000	0	1.0
2,000	0	0.89
4,000	0	0.85
8,000	0	0.66
12,000	0	0.44

11·8 The following data include the capacitance per unit area and the breakdown voltages of SiO_2—silicon structures in microcircuits for different oxide thicknesses. Plot the two curves on the same graph, identify the curves, and provide a suitable title.

curve 1		curve 2	
capacitance (picofarads/ sq mil.)	oxide thickness (angstrom units)	breakdown voltage (millivolts)	oxide thickness ($\mathring{A}$)
0.41	500	0.055	500
0.22	1,000	0.09	1,000
0.14	1,500	0.16	2,000
0.105	2,000	0.215	3,000
0.07	3,000		

11·9 Plot the family of curves for the 2N1490 NPN transistor having a base input on a common-emitter circuit. Curves for four base currents are shown below.

Collector characteristics

curve 1		curve 2	
25 base milliamps		10 base milliamps	
collector-to-emitter voltages	collector milliamps	collector-to-emitter voltages	collector milliamps
0	0	0	0
0.5	160	0.5	160
1.0	300	1.0	250
1.5	395	1.5	315
2.0	450	2.0	375
2.5	500	2.5	420
		3.0	470

Collector characteristics

curve 3		curve 4	
3 base milliamps		1 base milliamp	
collector-to-emitter voltages	collector milliamps	collector-to-emitter voltages	collector milliamps
0	0	0	0
0.5	110	0.5	90
1	190	1	112
2	270	2	125
3	320		Straight
4	352		to
		4.5	141

11·10   The table below provides data for four characteristic curves of a 2N2102 NPN transistor for an ambient temperature of 25° C. Plot the family of four curves for a common-emitter circuit having a base input. These are typical collector characteristics.

base current = 2MA		base current = 8	
collector-to-emitter volts	*collector milliamps*	*collector-to-emitter volts*	*collector milliamps*
0	0	0	0
0.3	100	0.3	200
1	140	1	250
2	175	3	365
4	195	4	402
10	210	6	450

base current = 14		base current = 22	
collector-to-emitter volts	*collector milliamps*	*collector-to-emitter voltages*	*collector milliamps*
0	0	0	0
0.3	200	0.3	200
0.6	300	0.6	300
1	330	1	375
2	402	2	482
4	502	2.2	500

11·11 On rectangular coordinate paper, plot the depreciation-cost curve and the maintenance-cost curve for

depreciation cost		maintenance cost	
cost/kwhr (cents)	*generator output (kw) (independent variable)*	*cost/kwhr (cents)*	*generator output (kw)*
3	3	2	2
2	10	1.5	10
1.8	20	1.35	20
1.6	80	1.2	40
1.5	160	1.25	80
		1.6	160

electronic heating equipment, using the data given below. A suitable title would be "Costs of 60 Hz Electrical Energy." Use different plotting symbols for each curve. Your instructor may want you to add the two curves graphically to get a total cost curve.

11·12 Plot the family of curves for a tube having the average plate characteristics shown above, right. Tube type: 6AL8. E = 6.3 volts.

$E_{c1} = 0$ curve for No. 1 grid		curve for $E_{c1} = -1$		curve for $E_{c1} = -2$		curve for $E_{c1} = -3$	
DC plate voltage (volts)	plate current (milli-amperes)	DC plate voltage (volts)	plate current (milli-amperes)	plate voltage (volts)	plate current (milli-amperes)	plate voltage (volts)	plate current (milli-amperes)
0	5	0	5	7	13	6	5
3	16	5	16	20	21	15	10
5	25	7	20	35	25	40	16
6	35	10	30	60	26.5	80	17
10	42	15	36	80	24	28	14
25	51	24	40				
50	55	45	44				
80	56	80	45				

11·13 The annual requirements for artwork for three types of circuit manufacturing have been rising and will continue to do so. The data shown below should be plotted on rectangular (arithmetic) graph paper to show past and estimated requirements through 1976.

Number of artwork layouts per year (in thousands)

year	printed wiring boards	thin-film circuits	silicon monolithic circuits
1968	6.0	4.6	1.5
1970	9.0	6.5	3.0
1972	14.2	9.0	5.0
1974	19.0	13.5	7.6
1976	23.0	20.6	11.6

The curves may be extrapolated for another year or two. Estimates were made by the economics division of Atlantic Electronics Company, January 30, 1972.

11·14   On semilogarithmic paper (three cycles) make a graph for the date given below. An appropriate title might be "Input Resistance versus Load Resistance of Transistor Circuit."

lead resistance (ohms, independent variable)	input resistance (ohms)
10K	1,020
50K	820
100K	700
300K	600
1 meg	560
10 meg	540

11·15   Construct a graph using multicycle semilogarithmic paper, for the phase response of an RC amplifier. Let the phase shift, below, be the dependent variable. Use the abbreviations Hz and kHz for cps, kc, etc.

response curve without feedback		response curve with feedback	
frequency (units as shown)	phase shift (degrees)	frequency (units as shown)	phase shift (degrees)
		10 cps	+ 130
10 cps	+ 150	40 cps	20
100 cps	50	100 cps	5
1 kc	0	1 kc	0
10 kc	−20	10 kc	0
100 kc	−120	100 kc	−20
1 Mc	−180	300 kc	−100
		1 Mc	−150

11·16   Make a graph showing the two curves, one with and one without feedback, for the frequency response of the transistor amplifier.

without feedback		with feedback	
frequency (cps)	gain (db)	frequency (cps)	gain (db)
10	48	10	18
30	54	35	20
100	56	100	21
10^4	55	10^3	21
10^5	43	4×10^4	21
3×10^5	35	9×10^4	22
(300,000)		2×10^5	20
		3×10^5	15

11·17   On polar coordinate paper, plot the dipole radiation pattern using the data listed below. There will be a major and two minor lobes. Do not number or circle points. This is a vertical plane pattern of a half-wave center-fed antenna. (Zero should be at bottom of chart.)

major lobe			minor lobe			minor lobe		
point	angle	amplitude	point	angle	amplitude	point	angle	amplitude
0	0°	0	0	0°	0	0	0°	0
1	155°	0.2	1	120°	0.10	1	215°	0.1
2	158°	0.4	2	124°	0.20	2	216°	0.2
3	162°	0.6	3	126°	0.24	3	218°	0.24
4	169°	0.8	4	130°	0.28	4	222°	0.28
5	180°	1.0	5	134°	0.30	5	226°	0.30
6	191°	0.8	6	138°	0.28	6	230°	0.28
7	198°	0.6	7	142°	0.24	7	234°	0.24
8	202°	0.4	8	144°	0.20	8	236°	0.20
9	205°	0.2	9	145°	0.10	9	240°	0.10
10	0°	0	10	0°	0	10	0°	0

11·18 The data listed below provide a vertical plane radiation pattern for a ¾ wave antenna. The 0 to 180° line represents the axis of the antenna. Make a polar plot, using an irregular (French) curve wherever possible. (It may be necessary to round off the tips of the lobes by careful freehand drawing.) Do not number points, or put plotting symbols around them. (You may put 0° at the left side of the graph if you wish.)

Point no.	0	1	2	3	4	5	6	7	8	9	10	11	12
Angle (°)	0	15	16	19	23	27	30	33	37	41	44	45	0
Amplitude	0	.2	.4	.6	.8	.96	1.0	.96	.8	.6	.4	.2	0

Point no.	0	1	2	3	4			0	1	2	3	4
Angle	0	60	64	70	0			0	75	81	86	0
Amplitude	0	.18	.3	.18	0			0	.18	.3	.18	0

Point no.	0	1	2	3	4			0	1	2	3	4
Angle	0	92	98	104	0			0	250	256	262	0
Amplitude	0	.18	.3	.18	0			0	.18	.3	.18	0

Point no.	0	1	2	3	4	5	6	7	8	9	10	11	12
Angle	0	135	136	139	143	147	150	153	157	161	164	165	0
Amplitude	0	.2	.4	.6	.8	.96	1.0	.96	.8	.6	.4	.2	0

The following graphical plots are of such a nature that each one yields a straight-line pattern if plotted on a certain type of graph paper. If the straight-line pattern can be achieved, then the student can write the equation for the line (and the data), using one of the equations listed in Sec. 11·12. Some experimentation of a trial-and-error nature will be required, in most cases, before the student will be able to draw the graph on the correct paper. Principles of graphical presentation should be followed, and the final graph should have a title and good line work and lettering, and the equation (if required by the instructor) should be shown prominently on the graph.

11·19 The following data provide a single-family transfer characteristic of a transistor which is part of an integrated circuit on a silicon wafer.

collector characteristic (I_c in milliamps)	gate voltage (V_G in volts)
−1	0.2
10	0.6
21	1.0
32	1.4
42	1.8

11·20 The following data are from test records on 2,409 integrated circuits. There are two curves, one for 90 percent and one for 95 percent upper-confidence limits.

90% curve		95% curve	
operating time (T in hrs)	failure rate (F in fails/ hr)	operating time (T in hrs)	failure rate (F in fails/ hr)
40,400	10^{-4} (1 in 10,000)	50,400	10^{-4}
10^5	4.4×10^{-4}	10^5	5.2×10^{-4}
10^6	4.2×10^{-5}	10^6	5.2×10^{-5}
4.2×10^6	10^{-6}	5.2×10^6	10^{-6}
10^7	4.0×10^{-6}	10^7	5×10^{-6}

11·21 The following data show the optimum frequency of different skin thickness of brass and iron. Plot frequency in megahertz (millions of cps).

frequency (CPS)	skin thickness in centimeters	
	brass	iron
100 thousand	0.007	0.06
200 thousand	0.0042	0.042
400 thousand	0.0026	0.03
600 thousand	0.0019	0.024
1 million	0.0013	0.019

11·22   The data listed below show the relationship of power
        loss to the armature voltage of a ⅓ HP electric
        motor.

loss (watts)	armature voltage (volts)
0.080	1.2
0.122	1.5
0.213	2
0.340	2.5
0.480	3
0.842	4

11·23   Below are data that show the resistance in ground
        connections according to how deep the grounding
        rod is placed in the soil.

R (ohms)	D (feet)
90	2
68	3
51	4
33	7
24	10
17	15
11.5	25

11·24   By stressing improvement in engineering and pro-
        duction services and cost reduction, United States
        industry can maintain its constant increase in pro-
        ductivity. The following data cover a 10-year period.
        Use 1950 for getting the intercept.

year	output per manhour (dollars)
1950	81
1955	93
1960	106
1965	121
1970	139

11·25   Failure to align the length standard when making linear measurements with a laser causes a cosine error which plots as follows:

misalignment angle X (minutes)	cosine error e_c (ppm)
2	2.0
3	4.5
4	8.0
6	18.0
10	51.0

11·26   The manufacturing tolerances of thin-film resistors, according to the present state of the art, are:

resistance value (ohms)	manufacturing tolerance (percent)
200	±0.29%
600	0.48
1K	0.78
1.4K	1.33

Appendixes

Appendix A

Glossary of electronics and electrical terms

Ampere The practical unit of current. One ampere will flow through a resistance of one ohm when a difference of potential of one volt is applied across its terminals.

Amplification The process of increasing the strength (current, power, or voltage) of a signal.

Amplifier A device used to increase the signal voltage, current, or power, generally composed of a transistor or vacuum tube and an associated circuit called a *stage*. It may contain several stages in order to obtain a desired gain.

Amplitude The maximum instantaneous value of an alternating voltage or current, measured in either the positive or the negative direction.

Anode A positive electrode or terminal; the plate of a vacuum tube. Sometimes the most positive electrode.

Attenuation The reduction in the strength of a signal.

Base One of (usually) three regions of a transistor. Also one of the terminals of a transistor. In some transistors the base acts much like the grid of an electron tube.

Bias Vacuum tube: the difference of potential between the control grid and the cathode. Transistor: the difference of potential between the base and emitter and the base and collector. Magnetic amplifier: the level of flux density in the magnetic amplifier core under no-signal condition.

Bus Bar A primary power-distribution point connected to the main power source.

Capacitor A device consisting of two conducting surfaces separated by an insulating material or dielectric such as air, paper, or mica. A capacitor stores electric energy, blocks the flow of direct current, and permits the flow of alternating current to a degree depending on the capacitance and frequency.

Cathode The electrode in a vacuum tube which is the source of electron emission; also a negative electrode.

Choke Coil A coil of low ohmic resistance and high impedance to alternating current.

Circuit Breaker An electromagnetic or thermal device that opens a circuit when the current in the circuit exceeds a predetermined amount. Circuit breakers can be reset.

Coaxial Cable A transmission line consisting of two conductors concentric with and insulated from each other.

Cold Cathode A cathode without a heater such as is found in fluorescent-type tubes.

Collector That region of a transistor that collects electrons, or the terminal that corresponds to the anode (plate) of the electron tube in the normal mode of operation.

Commutator The copper segments on the armature of a d-c motor or generator. It is cylindrical in shape and is used to pass power into or from the brushes. It is a switching device.

Conductance The ability of a material to conduct or carry an electric current. It is the reciprocal of the resistance of the material and is expressed in mhos.

Conductor Any material suitable for carrying electric current.

Cryogen (Cryogenic) A device that becomes a superconductor (has practically no resistance) at extremely cold temperatures. A circuit having such devices.

Current The rate of transfer of electricity. An amount of electricity. The basic unit is the ampere.

Deflecting Plate That part of a certain type of electron tube which deflects the electron beam within the tube itself.

Detection The process of separating the modulation component from the received signal.

Dielectric An insulator; a term that refers to the insulating material between the plates of a capacitor.

Diode Vacuum tube: a two-element tube that contains a cathode and plate. Semiconductor: a material of either germanium or silicon that is manufactured to allow current to flow in only one direction. Diodes are used as rectifiers and detectors.

Electrode A terminal used to emit, collect, or control electrons and ions; a terminal at which electric current passes from one medium into another.

Electron A negatively charged particle of matter.

Electron Emission The liberation of electrons from a body into space under the influence of heat, light, impact, chemical disintegration, or potential difference.

Emitter That part or element of a transistor that emits electrons; it corresponds to the cathode of an electron tube, in the most common form of operation.

Epitaxial A thin-film type of deposition for making certain devices in microcircuits. It involves a realignment of molecules and hence has a deeper significance than just thin-film manufacture.

Eyelet Eyelets are used on printed-circuit boards to make reliable connections from one side of the board to the other side.

Farad The unit of capacitance.

Feedback A transfer of energy from the output circuit of a device back to its input.

Filament An electrically heated wire that emits electrons or heats a cathode which then emits electrons.

Filter A combination of circuit elements designed to pass a definite range of frequencies, attenuating all others.

Frequency The number of complete cycles per second existing in any form of wave motion, such as the number of cycles per second of an alternating current.

Fuse A protective device inserted in series with a circuit. It contains a metal that will melt or break when current is increased beyond a specific value for a definite time period.

Gain The ratio of the output power, voltage, or current to the input power, voltage, or current, respectively.

Gate A device or circuit that makes an electronic circuit operative for a short time.

Grid A wire, usually in the form of a spiral, that controls the electron flow in a vacuum tube.

Grid Leak A high resistance connected across the grid capacitor or between the grid and the cathode to provide a d-c path from grid to cathode and to limit the accumulation of charge on the grid.

Ground A metallic connection with the earth to establish ground potential. Also, a common return to a point of zero potential. The chassis of a receiver or a transmitter is sometimes the common return and is therefore, the "ground" of the unit.

Henry The basic unit of inductance. The inductance of a circuit is one henry when a current variation of one ampere per second induces one volt. (The plural is henrys.)

Hole In semiconductors, the space in an atom left vacant by a departed electron. Holes flow in a direction opposite to that of electrons, are considered to be current carriers, and bear a positive charge.

Impedance The total opposition offered to the flow of an alternating current. It may consist of any combination of resistance, inductive reactance, and capacitive reactance.

Inductance The property of a circuit or two neighboring circuits which determines how much electromotive force will be induced in one circuit by a change of current in either circuit.

Inductor A circuit element designed so that its inductance is its most important electrical property; a coil.

Integrated Circuit A circuit in which several different types of devices such as resistors, capacitors, and transistors are made from a single piece of material such as a silicon chip, and then are connected to form a circuit.

Logic The arrangement of circuitry designed to accomplish certain objectives such as the addition of two signals. Used largely in computer circuits, but also used in other equipment such as automated machine tools and electric controls.

Magnetron A vacuum-tube oscillator containing two electrodes, in which the flow of electrons from cathode to anode is controlled by an externally applied magnetic field.

Modulation The process of varying the amplitude (amplitude modulation), the frequency (frequency modulation), or the phase (phase modulation) of a carrier wave in accordance with other signals in order to convey intelligence. The modulating signal may be an audio-frequency signal, video signal (as in television), electric pulses or tones to operate relays, etc.

Oscillator A circuit that is designed to generate an audio or radio frequency; a mode of amplification. Also the main device in such a circuit.

Oscilloscope An instrument for showing, visually, graphical representations of the waveforms encountered in electric circuits.

Pentode An electron tube with five electrodes or elements.

Permalloy An alloy of nickel and iron, having an abnormally high magnetic permeability.

Plate The principal anode (electrode) in an electron tube to which the electron stream is attracted. Also, one of the conductive electrodes in a capacitor or battery.

Potential The degree of electrification as referred to some standard such as the earth. The amount of work required to bring a unit quantity of electricity from infinity to the point in question.

Potentiometer A variable voltage divider; a resistor which has a variable contact arm so that any portion of the potential applied between its ends may be selected.

Power The rate of doing work or the rate of expending energy. The unit of electric power is the watt.

Raceway A raceway is any channel for enclosing conductors which is designed expressly and used solely for this purpose.

Rectifiers Devices used to change alternating current to unidirectional current. These may be vacuum tubes, semiconductors such as germanium and silicon, dry-disk rectifiers such as selenium and copper oxide, and also certain types of crystal.

Relay An electromechanical switching device that can be used as a remote control.

Resistance The opposition that a device or material offers to the flow of current. It determines the rate at which electric energy is converted into heat or radiant energy.

Resistor A circuit element whose chief characteristic is resistance; used to oppose the flow of current.

Resonance The condition existing in a circuit in which the inductive and capacitive reactances cancel each other.

Saturation The condition existing in any circuit when an

increase in the driving signal produces no further change in the resultant effect.

Semiconductor An element, such as germanium or silicon, from which transistors or diodes are made; the device itself.

Solenoid An electromagnetic coil that contains a movable plunger.

Synchronous Happening at the same time; having the same period and phase.

Tachometer An instrument for indicating revolutions per minute.

Thin-film Circuit A circuit made by depositing material on a substrate, such as glass or quartz, to form patterns that make devices such as resistors, and capacitors and their connections. The thickness of the film forming these devices is only a few microns (0.0001 cm).

Thyratron A gas-filled triode tube which is extensively used in electronic-control circuits.

Transducer A device that converts an input into a different type of output. Examples are microphones, speakers, lamps, vibrators, strain gages, and generators.

Transformer A device composed of two or more coils linked by magnetic lines of force; used to transfer energy from one circuit to another.

Triode A three-electrode vacuum tube containing a cathode, control grid, and plate. Also a three-region semiconductor.

Volt The unit of voltage, potential or emf. One volt will send a current of one ampere through a resistance of one ohm.

Voltage Used interchangeably with "potential." See *Potential*.

Watt The unit of electric power. In a d-c current a watt is equal to volts multiplied by amperes. In an a-c current the true power in watts is effective volts multiplied by effective amperes, then multiplied by the circuit power factor (1hp = 746 watts).

Electrical device reference designations

FROM MIL STD 16B

Alarm	DS	Antenna, aerial	E
Amplifier	A	Arrestor, lightning	E
Amplifier, rotating	G	Assembly	A
Annunciator	DS	Attenuator	AT

Audible signaling device	DS	Detector, crystal	CR
Autotransformer	T	Device, indicating	DS
Battery	BT	Dipole antenna	E
Bell	DS	Disconnecting device	S
Blower, fan, motor	B	Electron tube	V
Board, terminal	TB	Exciter	G
Breaker, circuit	CB	Fan	B
Buzzer	DS	Filter	FL
Cable	W	Fuse	F
Capacitor	C	Generator	G
Cell, aluminum or electrolytic	E	Handset	HS
Cell, light sensitive, photoemissive	V	Head, erasing, recording, reproducing	PU
Choke	L	Heater	HR
Circuit breaker	CB	Horn, howler	LS
Coil, hybrid	HY	Indicator	DS
Coil induction, relay tuning, operating	L	Inductor	L
		Instrument	M
Coil, repeating	T	Insulator	E
Computer	A	Interlock, mechanical	MP
Connector, receptacle, affixed to wall, chassis, panel	J	Interlock, safety, electrical	S
		Jack (see connector, receptacle, electrical)	
Connector, receptacle, affixed to end of cable, wire	P	Junction, coaxial or waveguide (tee or wye)	CP
Contact, electrical	E	Junction, hybrid	HY
Contactor, electrically operated	K	Key, switch	S
		Lamp, pilot or illuminating	DS
Contactor, mechanically or thermally operated	S	Lamp, signal	DS
		Line, delay	DL
Coupler, directional	DS	Loop antenna	E
Crystal detector	CR	Magnet	E
Crystal diode	CR	Meter	M
Crystal, piezoelectric	Y	Microphone	MK
Cutout, fuse	F	Mode transducer	MT
Cutout, thermal	S	Modulator	A
		Motor	B

Motor-generator	MG	Resistor	R
Mounting (not in electrical circuit and not in a socket)	MP	Rheostat	R
		Selenium cell	CR
		Shunt	R
Nameplate	N	Solenoid	L
Oscillator (excluding elect. tube used in oscillator)	Y	Speaker	LS
		Speed regulator	S
		Strip, terminal	TP
Oscilloscope	M	Subassembly	A
Pad	AT	Switch, mechanically or thermally operated	S
Part, miscellaneous	E		
Path, guided, transmission	W	Terminal board or strip	TB
Phototube	V	Test point	TP
Pickup, erasing, recording, or reproducing head	PU	Thermistor	RT
		Thermocouple	TC
		Thermostat	S
Plug (see connector)		Timer	M
Potentiometer	R	Transducer	MT
Power supply	A	Transformer	T
Receiver, telephone	HT	Transistor	Q
Receptacle (fixed connector)	J	Transmission path	W
		Tube, electron	V
Rectifier, crystal or metallic	CR	Varistor, assymmetrical	CR
Regulator, voltage (except electron tube)	VR	Varistor, symmetrical	RV
		Voltage regulator (except an electron tube)	VR
Relay, electrically operated contactor or switch)	K	Waveguide	W
		Winding	L
Repeater (telephone usage)	RP	Wire	W

The above list is not the complete list of devices shown in Mil Std 16B. It contains the more commonly used devices.

Examples of the use of the above in electrical drawings would be: $C201$, $L5$, $J1$, $J1(P301)$, $W1P2$, $T205$, $2A4R3$. The last example is explained as "the third resistor of the fourth subassembly of the second unit."

Abbreviations for drawings and technical publications

FROM MIL STD 12B

Adaptor	ADPT	Decibel	DB
Air circuit breaker	ACB	Diameter	DIA
Alternating current	AC	Direct current	DC
Alternating current volts	VAC	Double-pole double-throw	DPDT
Aluminum	AL	Double-pole single-throw	DPST
American Society of Mechanical Engineers	ASME	Drawing	DWG
		Dynamometer	DYNO
Ammeter	AM.	Dynamotor	DYNM
Ampere	AMP	Electric horsepower	EHP
Amplifier	AMPL	Electrolytic	ELECT.
Antenna	ANT.	Electronic Industries, Association	EIA
Armature	ARM.		
Arrestor	ARR	Engineer	ENGR
Attenuation, attenuator	ATTEN	Engineering	ENGRG
		Escutcheon	ESC
Audio frequency	AF	Exciter	EXC
Auto frequency control	AFC	Federal Communications Commission	FCC
Automatic gain control	AGC	Federal Power Commission	FPC
Battery	BAT	Field reversing	FFR
Beat-frequency oscillator	BFO	Flat head	FH
		Fluorescent	FLUOR
Bottom	BOT	Fuze	FZ
Cabinet	CAB.	Gage	GA
Capacitor	CAP.	Germanium	Ge
Cathode-ray tube	CRT	Grommet	GROM
Circuit	CKT	Guided missile	GM
Coaxial	COAX.	Heater	HTR
Collector	COLL	Heat treat	HT TR
Compress	COMP	High frequency	HF
Condenser	COND	High-frequency oscillator	HFO
Conductor	COND		
Conduit	CND	High voltage	HV
Counterclockwise	CCW	Horizon, horizontal	HORIZ
Cycles per second	HZ	Ignition	IGN

Indicator	IND	National Electrical	
Induction-capacitance	LC	Code	NEC
Induction	IND	Not to scale	NTS
Institute of Electrical		Oil circuit breaker	OCB
and Electronic Engi-		Oscillator	OSC
neers	IEEE	Oscilloscope	OSCP
Instrument	INST	Overload	OVLD
Intermediate frequency	IF	Pentode	PENT.
Junction box	JB	Phase	PH
Knockout	KO	Piezoelectric-crystal	
Kilocycle	KC	unit	CU
Kilohm	K	Polarity	PO
Kilovolt	KV	Potentiometer	POT.
Kilovolt ampere	KVA	Power supply	PWR SUP
Kilowatt	KW	Quick-opening device	QOD
Kilowatt-hour	KWH	Radar	RDR
Lighting	LTG	Radio	RAD.
Low frequency	LF	Radio frequency	RF
Low voltage	LV	Reactive volt-amp	VAR
Magnetic ampli-		Receptacle	RECP
fier	MAG AMPL	Reference	REF
Magnetic modula-		Reference line	REF L
tor	MAG MOD	Resistance	RES.
Manual	MAN	Resistance capaci-	
Master switch	MS	tance	RC
Medium frequency	MF	Resistance-capacitance	
Mega (10^6)	MEG	coupled	RC CPLD
Megacycle	MC	Resistor	RES.
Megohm	MEGO	Roundhead	RH
Meter	M	Saturable reactor	SR
Metering	MTRG	Schedule	SCH
Missile	MSL	Screw	SCR
Modify	MOD	Secondary	SEC
Modulator	MOD	Selector	SEL
Motor	MOT	Selenium	Se
Mounting	MTG	Series relay	SRE
Multiplex	MX	Servomechanism	SERVO.
Multivibrator	MVB	Shield	SHLD
National Aeronau-		Signal	SIG
tics and Space		Single-pole, double-	
Administration	NASA	throw	SPDT

Single-pole, single-throw	SPST	Transistor	Q
Slow operate (relay)	SO.	Transmitter	XMTR
Solenoid	SOL.	Tuning	TUN
Speaker	SPKR	Twisted	TW
Specification	SPEC	Ultrahigh frequency	UHF
Suppressor (ion)	SUPPR	Unfused	UNF
Switch	SW	Vacuum tube	VT
Switchboard	SWBD	Vacuum tube voltmeter	VTVM
Switchgear	SWGR	Var-hour meter	VRH
Synchronous	SYN	Variable frequency oscillator	VFO
Tachometer	TACH		
Technical circular	TC	Very high frequency	VHF
Technical manual	TM	Very low frequency	VLF
Telemeter	TLM	Video	VID
Terminal	TERM.	Video frequency	VDF
Test switch	TSW	Volt	V
Thermistor	TMTR	Voltage regulator	VR
Thermocouple	TC	Voltmeter	VM
Three-conductor	3/C	Volume	VOL
Three-phase	3 PH	Watt	W
Three-pole	3 P	Watt hour	WHR
Time delay	TD	Watt-hour meter	WHM
Transceiver	XCVR	Wattmeter	WM
Transformer	XFMR	Wire-wound	WW

Note: Some abbreviations are followed by a period. Most do not have a period.

These symbols are authorized for use in drawings for the Military, and in specifications. A similar list has been compiled by the American National Standards Institute.

The above is not the complete Military Standard. If in doubt about abbreviation, spell out the word.

Appendix B

The frequency spectrum

frequency	designation	abbrevi-ation	wavelength meters	centimeters
Below 3 kHz	Extremely low frequency	elf	30,000	
3–30 kHz	Very low frequency	vlf	30,000–10,000	
30–300 kHz	Low frequency	lf	10,000–1,000	
300–3000 kHz	Medium frequency	mf	1,000–100	
3–30 MHz	High frequency	hf	100–10	
30–300 MHz	Very high frequency	vhf	10–1	
300–3000 MHz	Ultrahigh frequency	uhf	1–0.1	
3000 MC–30	Superhigh frequency	shf		10–1
30–300 GHz	Extremely high frequency	ehf		1–0.1
300–3000 GHz	As yet unnamed			0.1–0.01

Note: The new IEEE and international standards utilize the term "Hertz" for cycles per second. Thus KC (old form) becomes kHz; MC (old form) becomes MHz; etc. kHz represents kilocycles; MHz, megacycles; and GHz, gigacycles (1 billion cps).

Width of copper foil conductors for printed circuits*

width of 0.00135–in. thick copper (inches)	width of 0.0027–in. thick copper (inches)	current capability (amperes)
1/64		1.5
1/32	1/64	2.5
1/16	1/32	3.5
	1/16	5.5
1/8		6.5
	1/8	8.0
	3/16	15.0

By permission of the Aerovox Corp.
*For a temperature rise of 40°C, the maximum recommended for lucite and nylon laminates.

Example: If the desired current is 3.5 amp, a conductor 0.0027 in. thick and 1/32 in. wide will carry it. Also, a 0.00135-in. thick copper foil that is 1/16 in. wide will carry 3.5 amp with a rise of 40°C or less.

Resistor color code

The colored bands around the body of a resistor indicate its value in ohms and its tolerance if there are four bands. As the drawing shows, the first band represents the first digit of the value and the second band the second digit. (The first band is near one edge and is often right at that edge of the device.) The third band represents the number by which the two digits are multiplied. A fourth band of gold or silver represents a tolerance of ±5 percent or ±10 percent respectively. If no fourth band is shown, the tolerance is assumed to be ±20 percent of the indicated value of resistance.

The physical size of a composition resistor is related to its wattage rating. Size increases progressively as the wattage rating increases. The diameters of 1/2-watt, 1-watt, and 2-watt resistors are approximately 1/8 in., 1/4 in., and $5/_{16}$ in. respectively.

CODE

COLOR	DIGIT		MULTIPLIER
	1 ST	2 ND	
BLACK	0	0	1
BROWN	1	1	10
RED	2	2	100
ORANGE	3	3	1,000
YELLOW	4	4	10,000
GREEN	5	5	100,000
BLUE	6	6	1,000,000
VIOLET	7	7	10,000,000
GRAY	8	8	100,000,000
WHITE	9	9	1,000,000,000
GOLD	—	—	.1
SILVER	—	—	.01

TOLERANCE
GOLD   ±   5%
SILVER   ± 10%
NO BAND   ± 20%

EXAMPLES

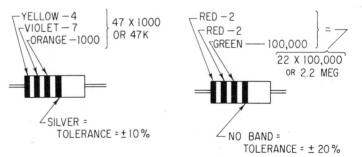

YELLOW — 4
VIOLET — 7   } 47 X 1000
ORANGE — 1000   OR 47K

SILVER =
TOLERANCE = ± 10%

RED — 2
RED — 2   } =
GREEN —— 100,000

22 X 100,000
OR 2.2 MEG

NO BAND =
TOLERANCE = ± 20%

Capacitor color codes

Of the several types and shapes of capacitors only the mica
and tubular ceramic types are color-coded. Color codes differ
somewhat among manufacturers, but the codes shown here
(page 354) apply to almost all the mica and tubular ceramic
capacitors that are in common use. These comply with the EIA
Standards.

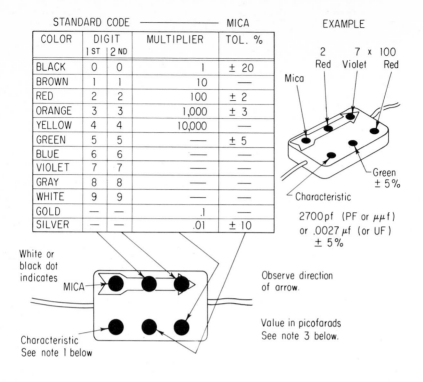

COLOR	DIGIT 1ST	DIGIT 2ND	MULTIPLIER	TOL. %
BLACK	0	0	1	± 20
BROWN	1	1	10	—
RED	2	2	100	± 2
ORANGE	3	3	1,000	± 3
YELLOW	4	4	10,000	—
GREEN	5	5	—	± 5
BLUE	6	6	—	—
VIOLET	7	7	—	—
GRAY	8	8	—	—
WHITE	9	9	—	—
GOLD	—	—	.1	—
SILVER	—	—	.01	± 10

EXAMPLE

2   7   x   100
Red   Violet   Red

Mica

Green
± 5%

Characteristic

2700 pf (PF or $\mu\mu$f)
or .0027 μf (or UF)
± 5%

White or
black dot
indicates
MICA

Characteristic
See note 1 below

Observe direction
of arrow.

Value in picofarads
See note 3 below.

STANDARD CODE ——————— TUBULAR

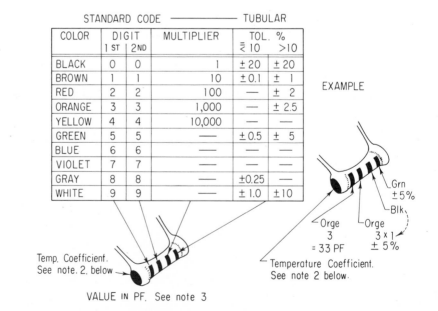

COLOR	DIGIT 1 ST	DIGIT 2ND	MULTIPLIER	TOL. % ⪝ 10	TOL. % >10
BLACK	0	0	1	± 20	± 20
BROWN	1	1	10	± 0.1	± 1
RED	2	2	100	—	± 2
ORANGE	3	3	1,000	—	± 2.5
YELLOW	4	4	10,000	—	—
GREEN	5	5	—	± 0.5	± 5
BLUE	6	6	—	—	—
VIOLET	7	7	—	—	—
GRAY	8	8	—	±0.25	—
WHITE	9	9	—	± 1.0	±10

EXAMPLE

Grn
± 5%

Blk

Orge
3 x 1
± 5%

Orge
3
= 33 PF

Temperature Coefficient.
See note 2 below.

Temp. Coefficient.
See note. 2. below

VALUE IN PF. See note 3

NOTES 1 The characteristic of a mica capacitor is the temperature coefficient, drift capacitance, and insulation resistance. This information is not usually needed to identify a capacitor, but if desired, it can be located in EIA (Electronic Industries Association) Standard RS-153.

2 The temperature coefficient of a capacitor is the predictable change in capacitance with temperature change and is expressed as parts per million per degree centigrade. (Refer to EIA Standard RS-198.)

3 Although the farad is the basic unit of capacitance, capacitor values are generally expressed in microfarads (μf or UF or .000001 farad) or picofarads (pf or PF, or $\mu\mu$f or UUF, or .000001 *microfarad*). Therefore, 1,000 picofarads = .001 microfarad, and 1,000,000 picofarads = 1 microfarad.

Color code for chassis wiring

color	*abbrev.*	*numerical code*	*circuit*
Black	BK	0	Grounds, grounded elements and returns
Brown	BR	1	Heaters of filaments off ground
Red	R	2	Power supply B-plus
Orange	O	3	Screen grids
Yellow	Y	4	Cathodes, emitters
Green	GN	5	Control grids, base
Blue	BL	6	Plates (anodes), collectors
Violet (or Purple)	V PR	7	Power supply, minus
Gray	GY	8	AC power lines
White	W	9	Miscellaneous, returns above or below ground, AVC, etc.

Source: Mil Std 122.

Circuit identification color code for industrial control wiring

circuit	*color*
Line, load, and control circuit at line voltage	Black
AC control circuit	Red
DC control circuit	Blue
Interlock panel control when energized from external force	Yellow
Equipment grounding conductor	Green
Grounded neutral conductor	White

Source: National Machine Tool Builders Association.

Transformer color codes

POWER TRANSFORMERS

1 Primary leads
  If tapped:
    Common   Black
    Tap   Black and yellow striped
    Finish   Black and red striped
2 High-voltage plate winding   Red
  Center tap   Red and yellow striped
3 Rectifier filament winding   Yellow
  Center tap   Green and yellow striped
4 Filament winding No. 1   Green
  Center tap   Green and yellow striped
5 Filament winding No. 2   Brown
  Center tap   Brown and yellow striped
6 Filament winding No. 3   Slate
  Center tap   Slate and yellow striped

AUDIO TRANSFORMERS

Blue	Plate (finish) lead of primary
Red	B + lead (no difference with center tap)
Brown	Plate (start) lead on center-tapped primaries. (Blue may be used if polarity is not important.)
Green	Grid (finish) lead to secondary.
Black	Grid return (no difference with center tap).
Yellow	Grid (start) lead on center-tapped secondaries. (Green may be used if polarity is not important.)

I-F TRANSFORMERS

Blue	Plate lead
Red	B + lead
Green	Grid (or diode) lead
Black	Grid (or diode) return

Metric conversion table

millimeters to inches				inches to millimeters				inches (decimals) to millimeters			
mm	in	mm	in	in	mm	in	mm	in	mm	in	mm
1 = 0.0394		17 = 0.6693		$\frac{1}{32}$ =	0.794	$\frac{17}{32}$ =	13.493	0.01 = 0.254		0.25 =	6.350
2 = 0.0787		18 = 0.7087		$\frac{1}{16}$ =	1.587	$\frac{9}{16}$ =	14.287	0.02 = 0.508		0.26 =	6.604
3 = 0.1181		19 = 0.7480		$\frac{3}{32}$ =	2.381	$\frac{19}{32}$ =	15.081	0.03 = 0.762		0.28 =	7.112
4 = 0.1575		20 = 0.7874		$\frac{1}{8}$ =	3.175	$\frac{5}{8}$ =	15.875	0.04 = 1.016		0.30 =	7.620
5 = 0.1969		21 = 0.8268		$\frac{5}{32}$ =	3.968	$\frac{21}{32}$ =	16.668	0.05 = 1.270		0.32 =	8.128
6 = 0.2362		22 = 0.8662		$\frac{3}{16}$ =	4.762	$\frac{11}{16}$ =	17.462	0.06 = 1.524		0.34 =	8.636
7 = 0.2756		23 = 0.9055		$\frac{7}{32}$ =	5.556	$\frac{23}{32}$ =	18.256	0.07 = 1.778		0.36 =	9.144
8 = 0.3150		24 = 0.9449		$\frac{1}{4}$ =	6.350	$\frac{3}{4}$ =	19.050	0.08 = 2.032		0.38 =	9.652
9 = 0.3543		25 = 0.9843		$\frac{9}{32}$ =	7.144	$\frac{25}{32}$ =	19.843	0.09 = 2.286		0.40 =	10.160
10 = 0.3937		26 = 1.0236		$\frac{5}{16}$ =	7.937	$\frac{13}{16}$ =	20.637	0.10 = 2.540		0.50 =	12.699
11 = 0.4331		27 = 1.0630		$\frac{11}{32}$ =	8.731	$\frac{27}{32}$ =	21.431	0.12 = 3.048		0.60 =	15.240
12 = 0.4724		28 = 1.1024		$\frac{3}{8}$ =	9.525	$\frac{7}{8}$ =	22.225	0.14 = 3.556		0.70 =	17.780
13 = 0.5118		29 = 1.1418		$\frac{13}{32}$ =	10.319	$\frac{29}{32}$ =	23.018	0.16 = 4.064		0.80 =	20.320
14 = 0.5512		30 = 1.1811		$\frac{7}{16}$ =	11.112	$\frac{15}{16}$ =	23.812	0.18 = 4.572		0.90 =	22.860
15 = 0.5906		50 = 1.9685		$\frac{15}{32}$ =	11.906	1 =	25.400	0.20 = 5.080		1.00 =	25.400
16 = 0.6299		100 = 3.9370		$\frac{1}{2}$ =	12.699	12 =	304.800	0.22 = 5.588		2.00 =	50.800

Control device designations[1]

BR	Brake relay	CRM	Control relay master
CR	Control relay	D	Down
CRH	Control relay manual	DISC	Disconnect switch
ET	Electron tube	OL	Overload relay
FLS	Flow switch	PB	Push button
FS	Float switch	R	Reverse
IOL	Instantaneous over-load	RH	Rheostat
		S	Switch
LS	Limit switch	SOL	Solenoid
M	Motor starter	SS	Selector switch
MB	Magnetic brake	T	Transformer
MC	Magnetic clutch	TR	Time delay relay
MN	Manual	X	Reactor

[1] The Joint Industry Committee (JIC), *"Electrical Standards for Industrial Equipment."*

Examples of relay designations:

General use	CR, 1 CR, 2 CR, etc.
Timers	TR, 1 TR, 2 TR
Overload	OL, 1 OL, 2 OL

Approximate radii for aluminum alloys for 90° cold bend*

alloy and temper	radii for various thicknesses in terms of thickness t				
	$\frac{1}{64}$ in.	$\frac{1}{32}$ in.	$\frac{1}{16}$ in.	$\frac{1}{8}$ in.	$\frac{3}{16}$ in.
1100-0	0	0	0	0	0
1100-H12	0	0	0	0	0-1
1100-H14	0	0	0	0	0-1
1100-H16	0	0	0-1	½-1½	1-2
1100-H18	0-1	½-1½	1-2	1½-3	2-4
Alclad 2014-0	0	0-1	0-1	0-1	0-1
Alclad 2014-T3	1-2	1½-3	2-4	3-5	4-6
Alclad 2014-T4	1-2	1½-3	2-4	3-5	4-6
Alclad 2014-T6	2-4	3-5	3-5	4-6	5-7
2024-0	0	0-1	0-1	0-1	0-1
2024-T3	1½-3	2-4	3-5	4-6	4-6
2024-T36	2-4	3-5	4-6	5-7	5-7
2024-T4	1½-3	2-4	3-5	4-6	4-6
2024-T81	3½-5	4½-6	5-7	6½-8	7-9
2024-T86	4-5½	5-7	6-8	7-10	8-11
2219-T31			½-1½	1-1½	1-2
2219-T37			½-2	1½-3	2-3½
2219-T81			2-4	2½-4½	3-5
2219-T87			2-4	3-5	4-6
3003-0	0	0	0	0	0
3003-H12	0	0	0	0	0-1
3003-H14	0	0	0	0-1	0-1
3003-H16	0-1	0-1		1-2	1½-3
3003-H18	½-1½	1-2	1½-3	2-4	3-5

*"Alcoa Aluminum Handbook," by permission of Aluminum Company of America.

Thickness of wire and metal sheet gages (inches)

gage number	American or Brown and Sharpe gage 1*	United States Standard gage 2†	gage number	American or Brown and Sharpe gage 1*	United States Standard gage 2†
0	0.3249	0.3125	16	0.0508	0.0625
1	0.2893	0.2813	17	0.0453	0.0563
2	0.2576	0.2656	18	0.0403	0.0500
3	0.2294	0.2500	19	0.0359	0.0438
4	0.2043	0.2344	20	0.0320	0.0375
5	0.1819	0.2188	21	0.0285	0.0344
6	0.1620	0.2031	22	0.0253	0.0313
7	0.1443	0.1875	23	0.0226	0.0281
8	0.1285	0.1719	24	0.0201	0.0250
9	0.1144	0.1563	25	0.0179	0.0219
10	0.1019	0.1406	26	0.0159	0.0188
11	0.0907	0.1250	27	0.0142	0.0172
12	0.0808	0.1094	28	0.0126	0.0156
13	0.0720	0.0938	29	0.0113	0.0141
14	0.0641	0.0781	30	0.0100	0.0125
15	0.0571	0.0703	31	0.0089	0.0109

*For aluminum sheet, rod, and wire. Also for copper wire, and brass, alloy and nickel silver wire and sheet.
†For steel, nickel, and Monel metal sheets.

Minimum radius of conduit* (inches)

size of conduit (inches)	national machine tool builders' association	NEC—for conductors without lead sheath	NEC—for conductors with lead sheath
½	4	4	6
¾	4½	5	8
1	5¾	6	11
1¼	7	8	14
1½	8¼	10	16
2	9½	12	21
2½	10½	15	25
3	13	18	31
3½	15	21	36
4	16	24	40
5	24	30	50
6	30	36	60

*The radius is that of the inner edge. Fittings shall be threaded unless structural difficulties prevent assembly. A run of conduit shall not contain more than the equivalent of 4 quarter bends (360°) total.

Wire gage sizes and current capacity

AWG no.	nominal diameter (inches)		ampere rating*
	bare	enameled	
10	0.1019	0.1039	30
12	0.0808	0.0827	20
14	0.0641	0.0661	15
16	0.0508	0.0526	10
18	0.0403	0.0419	7
20	0.0320	0.0335	5
22	0.0253	0.0267	3
24	0.0201	0.0213	
26	0.0159	0.0170	
28	0.0126	0.0135	
30	0.0100	0.0108	
32	0.0080	0.0087	
34	0.0063	0.0069	
36	0.0050	0.0055	
38	0.0040	0.0044	
40	0.0031	0.0035	

*From National Machine Tool Builders' Association.
Note the logical relationship between wires wherein every third conductor is half the size, going down the scale. NAS 729 offers a scale for flat conductors between AWG 22 and AWG 32.

Decimal equivalents of fractions*

fraction	equivalent	fraction	equivalent	fraction	equivalent	fraction	equivalent
1/64	0.0156	17/64	0.2656	33/64	0.5156	49/64	0.7656
1/32	0.0312	9/32	0.2812	17/32	0.5312	25/32	0.7812
3/64	0.0468	19/64	0.2968	35/64	0.5468	51/64	0.7968
1/16	0.0625	5/16	0.3125	9/16	0.5625	13/16	0.8125
5/64	0.0781	21/64	0.3281	37/64	0.5781	53/64	0.8281
3/32	0.0937	11/32	0.3437	19/32	0.5937	27/32	0.8437
7/64	0.1093	23/64	0.3593	39/64	0.6093	55/64	0.8593
1/8	0.1250	3/8	0.3750	5/8	0.6250	7/8	0.8750
9/64	0.1406	25/64	0.3906	41/64	0.6406	57/64	0.8906
5/32	0.1562	13/32	0.4062	21/32	0.6562	29/32	0.9062
11/64	0.1718	27/64	0.4218	43/64	0.6718	59/64	0.9218
3/16	0.1875	7/16	0.4375	11/16	0.6875	15/16	0.9375
13/64	0.2031	29/64	0.4531	45/64	0.7031	61/64	0.9531
7/32	0.2187	15/32	0.4687	23/32	0.7187	31/32	0.9687
15/64	0.2343	31/64	0.4843	47/64	0.7343	63/64	0.9843
1/4	0.2500	1/2	0.5000	3/4	0.7500	1	1.000

*See page 359 for metric equivalents.

Binary and other number systems*

decimal	conventional binary	reflected binary	excess-3 code	8421 code	octal code	hexi-decimal
0	0000	0000	0011	0000	0	0
1	0001	0001	0100	0001	1	1
2	0010	0011	0101	0010	2	2
3	0011	0010	0110	0011	3	3
4	0100	0110	0111	0100	4	4
5	0101	0111	1000	0101	5	5
6	0110	0101	1001	0110	6	6
7	0111	0100	1010	0111	7	7
8	1000	1100	1011	1000	10	8
9	1001	1101	1100	1001	11	9
10	1010	1111	1101	0001 0000	12	F
11	1011	1110	1110	0001 0001	13	G
12	1100	1010	1111	0001 0010	14	J
13	1101	1011	10000	0001 0011	15	K
14	1110	1001	10001	0001 0100	16	Q
15	1111	1000	10010	0001 0101	17	W
16	10000	11000	10011	0001 0110	20	10

*These number systems or codes are typical of the internal number systems in digital computers.

NOTES

1 The conventional or "straight" binary is the fundamental binary code in which the base is 2. Thus $2^1 = 10$ (binary), $2^2 = 100$, $2^3 = 1000$, and so on.

2 The reflected binary, which is also called the "Gray" code, has two noteworthy characteristics. First, it is reflective, as for example: 0011 (for 2) is reflective of 1100 (for 8). Also, each successive number represents a change of only one character (bit) from the preceding number; thus 5 (0111) is different from 4 (0110) only in the fourth character.

3 The excess-3 code is advanced by three binary numbers; 0011 (representing zero) is the same as 0011 (representing 3) in the conventional binary code.

4 The octal code has the number 8 as its base. Thus 8 becomes 10, and 16 becomes 20.

5 The 8421 code expresses each decimal by its 4-bit binary equivalent. The number 578 is coded as 0101 0111 1000.

ADDITION OF BINARY NUMBERS

Decimal No. 4   100
Decimal No. 2   + 10
                   ──────
                     110 = decimal No. 6

This presents no carry-over problem.

Decimal No. 4   100
Decimal No. 6   + 110
                   ──────
                    1010 = decimal No. 10

This requires a carry-over in that column which is third from the right.

MULTIPLICATION OF BINARY NUMBERS

Decimal No. 10   1010
Decimal No. 14   × 1110
                     ────────
                           0
                        1010
                        1010
                        1010
                     ────────
                    10001100 = decimal No. 140

The addition part, of course, follows the rules of binary addition.

Appendix C

SYMBOLS FOR ELECTRICAL AND ELECTRONIC DEVICES

This list is abridged from ANSI Y32.2, "Graphical Symbols for Electrical and Electronics Diagrams," which is nearly identical to Mil Std 15-1. Symbols for use on logic diagrams are abridged from ANSI Y32.14, "Graphical Symbols for Logic Diagrams."[1]

Certain specialized build-up applications of basic symbols are omitted. Where the American Standard shows both single-line and complete symbol equivalents, this chart shows only the single-line symbol. Consult the basic Standard for complete symbols in these cases.

[1]ANSI Y32.2, "Graphical Symbols for Electrical and Electronics Diagrams," and ANSI Y32.14, "Graphical Symbols for Logic Diagrams," are published by the American National Standards Institute, Inc., 1430 Broadway, New York, N.Y. 10018, and are sponsored by the Institute of Electrical and Electronics Engineers and the American Society of Mechanical Engineers.

ADJUSTABLE

preset, continuous,
non-linear

AMPLIFIER

general

with two inputs

with two outputs

with adjustable gain

with associated attenuator

with associated power
supply

with external feedback
path

AMPLIFIER LETTER COMBINATIONS

(May be used with amplifier
symbols if needed for ex-
planation.)

BDG bridging
BST booster
CMP compression
DC direct current
EXP expansion
LIM limiting
MON monitoring
PGM program
PRE preliminary
PWR power
TRQ torque

ANTENNA

general

dipole

loop

OR

loop antenna (alternate
symbol)

counterpoise, antenna

ARRESTER, LIGHTNING

general

carbon block

electrolytic or aluminum
cell

horn gap

protective gap

sphere gap

valve or film element

multigap

ATTENUATOR, FIXED

See also **PAD** (same
symbols as variable attenuator without adjustment
arrow.)

ATTENUATOR VARIABLE

general

balanced

unbalanced

AUDIBLE SIGNALING DEVICE

bell

buzzer

loudspeaker

LOUDSPEAKER LETTER COMBINATIONS

* **HN** horn, electrical
* **HW** howler
* **LS** loudspeaker
* **SN** siren
† **EM** electromagnetic with
moving coil
† **EMN** electromagnetic,
moving coil and neutra-
lized winding
† **MG** magnetic armature
† **PM** permanent magnet

(Asterisk (*) and dagger (†)
are not part of symbol.)

sounder, telegraph

BATTERY

one cell

multicell

multicell with taps

multicell with adjustable
tap

CAPACITOR

general

polarized

adjustable or variable

adjustable or variable with mechanical linkage

continuously adjustable or variable differential

phase shifter

split stator

feed-through

CELL, PHOTOSENSITIVE

asymmetrical photoconductive transducer

symmetrical photoconductive transducer

photovoltaic transducer

CIRCUIT BREAKER

general

CIRCUIT ELEMENT

general

LETTER COMBINATIONS FOR CIRCUIT ELEMENTS

(* Asterisk not part of symbol.)
CB circuit breaker
DIAL telephone dial
EQ equalizer
FAX facsimile set
FL filter
FL-BE filter, band elimination
FL-BP filter, band pass
FL-HP filter, high pass
FL-LP filter, low-pass
NET network
PS power supply
RU reproducing unit
RG recording unit
TEL telephone station
TPR teleprinter
TTY teletypewriter

ADDITIONAL LETTER COMBINATIONS

(Specific graphical symbols preferred.)

AR amplifier
AT attenuator
C capacitor
HS handset
I indicating lamp
L inductor
LS loudspeaker
J jack
MIC microphone
OSC oscillator
PAD pad
P plug
HT receiver, headset
K relay
R resistor
S switch
T transformer
WR wall receptacle

GROUND

earth ground

chassis connection

common connections

*

(Identifying marks to denote points tied together shall replace (*) asterisks.)

CLUTCH; BRAKE

clutch disengaged when operating means deenergized

OR

clutch engaged when operating means deenergized

OR

brake applied when operating means energized

OR

brake released when operating means energized

OR

COIL, OPERATING (RELAY)

(Replace asterisk (*) with device designation.)

OR   OR

dot shows inner end of winding

OR

CONNECTION, MECHANICAL (INTERLOCK)

with fulcrum

CONNECTOR

female contact

male contact

separable connectors
(engaged)

OR

separable connectors
(alternate symbol)

coaxial connector with outside conductor carried through

two-conductor switchboard jack

two-conductor switchboard plug

female contact
(convenience outlets and mating connectors)

male contact (convenience outlets and mating connectors)

two-conductor nonpolarized connector with female contacts

two-conductor polarized connector with male contacts

WAVEGUIDE FLANGES

mated (general)

plain (rectangular waveguide)

choke (rectangular waveguide)

CONTACT, ELECTRICAL

fixed contact for jack, key or relay
→ OR ⊸ OR ⇁

fixed contact for switch
○ OR →

fixed contact for momentary switch
↴

sleeve
▯ OR ▯ OR ⌐

moving contact, adjustable
→ OR →

moving contact, locking
○⌒

moving contact, nonlocking
○—

segment, bridging contact
▱ OR ⬭

vibrator reed
○—▭

vibrator split reed
○—-▭

rotating contact
—⊙—

closed contact, break
 OR OR ○⊸○

open contact, make
OR ○⊸ OR ○⊸

transfer
OR ○⊸ OR

make-before-break

open contact with time-closing or time-delay-closing
TC ╧ OR ╧ TDC

closed contact with time-opening or time-delay-opening
TO OR ╪ TDO

time-sequential-closing
 OR

CORE

air core

NO SYMBOL

magnetic core of inductor or transformer
═

core of magnet
▭

COUNTER, ELECTROMECHANICAL
⊏▯ ⟋○

COUPLER, DIRECTIONAL

general
✕

E-plane aperture coupling, 30-db loss
✕ Ⓔ 30DB

loop coupling, 30-db loss
✕⌣ 30DB

probe coupling, 30-db loss
✕ ▮ 30DB

resistance coupling, 30-db loss
 30DB

COUPLING

(by aperture of less than waveguide size)
⊛

(Replace asterisk (*) by E, H or HE depending upon type of coupling to guided transmission path.)

DELAY FUNCTION

general
⊏▭ * ▭

tapped delay

(Replace asterisk (*) with value of delay.)

DIRECTION OF FLOW

one way
→
OR
→

both ways

DISCONTINUITY

equivalent series element

capacitive reactance

inductive reactance

inductance-capacitance
circuit, infinite reactance
at resonance

inductance-capacitance
circuit, zero reactance
at resonance

resistance

equivalent shunt element

capacitive susceptance

conductance

inductive susceptance

inductance-capacitance
circuit with infinite
susceptance at resonance

inductance-capacitance
circuit with zero
susceptance at resonance

ELECTRON TUBE

directly heated cathode,
heater

indirectly heated cathode

cold cathode
(including ionically heated
cathode)

photocathode

pool cathode

ionically heated cathode
with supplementary
heating

grid

deflecting electrode

ignitor

excitor

anode or plate

target or x-ray anode

dynode

composite anode-
photocathode

composite anode-cold
cathode

composite anode-ionically
heated cathode with
supplementary heating

shield, within envelope
and connected to a
terminal

outside envelope of x-ray
tube

coupling by loop

resonator, cavity type—
single-cavity envelope with
grid electrodes

resonator—double cavity
envelope with grid
electrodes

multicavity magnetron
anode and envelope

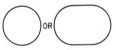

envelope

split envelope

gas-filled envelope

basing orientation, tubes
with keyed bases

KEY →

basing, tubes with
bayonets, bosses or other
reference points

base terminals

SMALL PIN

LARGE PIN

envelope terminals

RIGID TERMINAL

FLEXIBLE LEAD

triode with directly heated cathode and envelope connection to base terminal

pentode

twin triode equipotential cathode

cold-cathode voltage regulator

vacuum phototube

multiplier phototube

cathode-ray tube, electrostatic deflection

cathode-ray tube, magnetic deflection

mercury-pool tube with ignitor and control grid

mercury-pool tube with exciter, control grid and holding anode

single-anode pool-type vapor rectifier with ignitor

six-anode metal-tank pool-type rectifier with exciter

resonant magnetron with coaxial output

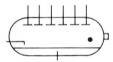

resonant magnetron with permanent magnet

PM

transit-time magnetron

tunable magnetron

reflex klystron, integral cavity

double-cavity klystron, integral cavity

transmit-receive (t-r) tube

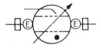

x-ray tube with directly heated cathode and focusing grid

x-ray tube with control grid

x-ray tube with grounded shield

double-focus x-ray tube with rotating anode

x-ray tube with multiple accelerating electrode

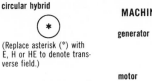

FUSE

general

OR

OR

high-voltage fuse

OR

high-voltage fuse, oil

OR

GOVERNOR

HALL GENERATOR

HANDSET

HYBRID

general

HYB

hybrid junction

H

E

circular hybrid

*

(Replace asterisk (*) with E, H or HE to denote transverse field.)

INDUCTOR

general

OR

magnetic-core inductor

tapped inductor

adjustable inductor

continuously adjustable inductor

saturable-core inductor
(reactor)

DC WINDING

KEY, TELEGRAPH

LAMP

ballast tube

B

fluorescent lamp, two-terminal

fluorescent lamp, four terminal

cold-cathode glow lamp, a-c type

cold-cathode glow lamp, d-c type

incandescent lamp

MACHINE, ROTATING

generator

GEN

motor

1-phase

3-phase wye grounded

3-phase wye ungrounded

3-phase delta

MAGNET, PERMANENT

PM

METER

*

METER LETTER COMBINATIONS

(Replace asterisk (*) with proper letter combination.)

A ammeter
AH ampere-hour
CMA contact-making
or breaking ammeter
CMC contact-making
or breaking clock
CMV contact-making
or breaking voltmeter
CRO cathode-ray
oscilloscope
DB decibel meter
DBM decibels referred to
one milliwatt
DM demand meter
DTR demand-totalizing relay
F frequency meter
G galvanometer
GD ground detector
I indicating
INT integrating
μ**A** or **UA** microammeter
MA milliammeter
NM noise meter
OHM ohmmeter
OP oil pressure
OSCG oscillograph, string
PH phasemeter

PI position indicator
PF power factor
RD recording demand meter
REC recording
RF reactive factor
SY synchroscope
T temperature
THC thermal converter
TLM telemeter
TT total time
V voltmeter
VA volt-ammeter
VAR varmeter
VARH varhour meter
VI volume indicating
VU standard volume
 indicating
W wattmeter
WH watthour meter

MICROPHONE

MODE SUPPRESSION

MODE TRANSDUCER

MOTION, MECHANICAL

translation, one direction

translation, both directions

rotation, one direction

rotation, both directions

NETWORK

NET

OSCILLATOR

PAD, (UNIDIRECTIONAL ISOLATOR)

PATH, TRANSMISSION

general

wire

two conductors

air or space path

dielectric path other than air

DIEL

crossing of conductors not connected

junction

junction of connected paths, conductors or wires

OR

OR ONLY IF REQUIRED BY SPACE LIMITATION

shielded single-conductor cable

coaxial cable

two-conductor cable

shielded two-conductor cable with shield grounded

grouping of leads

OR

OR

OR

OR

alternate or conditional wiring

associated or future wiring

associated or future equipment
(amplifier shown)

circular waveguide

rectangular waveguide

PHASE SHIFTER

general

adjustable

PICKUP HEAD

general

recording

playback

erasing

writing, reading and erasing

stereo

PIEZOELECTRIC CRYSTAL

POLARITY

positive

+

negative

−

RECEIVER, TELEPHONE

general

headset

RECTIFIER

(Represents any method of rectification such as electron tube, solid-state device, electrochemical device, etc.)

general

controlled

bridge type

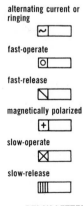

RELAY

alternating current or ringing

fast-operate

fast-release

magnetically polarized

slow-operate

slow-release

RELAY LETTER COMBINATIONS

(Not required with specific symbol.)

AC alternating current

D differential
DB double biased
DP dashpot
EP electrically polarized
FO fast operate
FR fast release
MG marginal
NB no bias
NR nonreactive
P magnetically polarized
SA slow operate and slow release
SO slow operate
SR slow release
SW sandwich wound

RESISTOR

general

—W— OR —[*]—

tapped resistor

—Λ— OR —[*]—

tapped resistor with adjustable contact

—W— OR —[*]—

adjustable or continuously adjustable

—W— OR —[*]—

instrument or relay shunt

—[o o]—

nonlinear resistor

—W— OR —[*]—

symmetrical varistor

—W— OR —[*]—
  V   V

OR —▷|—

(Replace asterisks (*) with identification of symbol.)

RESONATOR, TUNED CAVITY

ROTARY JOINT

general (Replace asterisk (*) with transmission-path recognition symbol.)

—[(*)]—

coaxial in rectangular waveguide

—[(⊕)]—

circular in rectangular waveguide

—[(⊙)]—

SEMICONDUCTOR DEVICES

semiconductor region with one ohmic connection

semiconductor region with plurality of ohmic connections

—| OR —⊤ OR ⊤⊤

rectifying junction, P on N region

—▼— OR —▼—

rectifying junction, N on P region

—▲— OR —▲—

emitter, P on N region

plurality of P emitters on N region

emitter, N on P region

plurality of N emitters on P region

collector

plurality of collectors

transition between regions of dissimilar conductivity

intrinsic region between regions of dissimilar conductivity

intrinsic region between regions of similar conductivity

intrinsic region between collector and region of dissimilar conductivity

intrinsic region between collector and region of similar conductivity

light dependence

temperature dependence
t°

capacitive device

tunneling device
]

breakdown device
⌐

PNP transistor (actual device and construction of symbol)

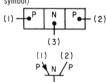

(1)   (2)

(3)

PNINIP device (actual device and construction of symbol)

(4)(1)(5)(2)(6)   (9)

(7)(8)   (3)

(1) (2)   I   I (3)

(4)(5)(6)(7)(8)(9)

semiconductor diode
(also: rectifier)

OR

OR

capacitive diode (also: Varicap, varactor, reactance diode, parametric diode)

OR

breakdown diode, unidirectional
(also: backward diode, avalanche diode, voltage regulator diode, zener diode, voltage reference diode)

OR

breakdown diode, bidirectional and backward diode
(also: bipolar voltage limiter)

OR

tunnel diode (also esaki diode)

OR

temperature dependent diode

t°   OR   t°

photodiode (also: solar cell)

OR

semiconductor diode, PNPN switch (also: Shockley diode, four-layer diode)

   OR

PNP transistor (also: junction, point-contact, mesa, epitaxial, planar, surface-barrier)

PNP transistor with one electrode connected to envelope

NPN transistor (see other names under PNP transistor)

unijunction transistor, N-type base (also: double-base diode, filamentary transistor)

unijunction transistor, P-type base (see other names above)

field-effect transistor N-type base

OR

field-effect transistor, P-type base

OR

semiconductor triode, PNPN switch (also: controlled rectifier)

semiconductor triode, NPNP switch (also: controlled rectifier)

NPN transistor with transverse-biased base

   OR

PNP transistor with ohmic connection to intrinsic region

NPN transistor with ohmic connection to intrinsic region

PNN transistor with ohmic connection to intrinsic region

NPIP transistor with ohmic connection to intrinsic region

SHIELD

SQUIB

explosive

igniter

sensing link

SWITCH

single-throw

double-throw

double-pole, double-throw with terminals shown

with horn gap

knife switch

push button, circuit closing (make)

push button, circuit opening (break)

nonlocking; momentary or spring return—circuit closing (make)

OR

nonlocking; momentary or spring return—circuit opening (break)

OR

nonlocking; momentary or spring return—transfer

OR

locking—circuit closing (make)

OR

locking—circuit opening (break)

OR

locking—transfer, three-position

OFF

selector switch

OR

selector, shorting during contact transfer

OR

wafer (example shown: 3-pole, 3-circuit with 2 nonshorting and 1 shorting moving contacts)

safety interlock—circuit opening

safety interlock—circuit closing

SWITCHING FUNCTION

conducting, closed contact (break)

nonconducting, open contact (make)

transfer

 OR

SYNCHRO

general

SYNCHRO LETTER COMBINATIONS

CDX control-differential transmitter

CT control transformer

CX control transmitter

TDR torque-differential receiver

TDX torque-differential transmitter

TR torque receiver

TX torque transmitter

RS resolver

B outer winding rotable in bearings

TERMINATION

cable

open circuit

short circuit

movable short

terminating series capacitor, path open

terminating series capacitor, path shorted

terminating series inductor, path open

terminating series inductor, path shorted

terminating resistor

series resistor, path open

series resistor, path shorted

THERMAL ELEMENT

actuating device

OR

thermal cutout

OR

thermal relay

OR

OR OR

OR

OR

thermostat (operates on rising temperature), **with break contact**

OR

thermostat with make
contact

thermostat with integral
heater and transfer
contacts

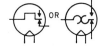

THERMISTOR

general

with integral heater

THERMOCOUPLE

general

with integral heater
internally connected

heater

with integral insulated
heater

heater

semiconductor
thermocouple, temperature
measuring

semiconductor
thermocouple, current
measuring

TRANSFORMER

general

OR

transformer with polarity
marks (instantaneous
current in to instantaneous
current out)

OR

one winding with
adjustable inductance

each winding with
adjustable inductance

adjustable mutual inductor

adjustable transformer

current transformer with
polarity marking

OR

bushing type current
transformer

potential transformer

OR

TRANSFORMER
CONNECTION
WINDING

3-phase 3-wire Delta or
mesh

3-phase 3-wire Delta
grounded

3-phase open Delta
grounded at common
point

3-phase wye or star
ungrounded

TERMINAL BOARD
OR STRIP

VIBRATOR

shunt drive

separate drive

VISUAL SIGNALING
DEVICE

annunciator, general

annunciator drop or
signal, shutter type

annunciator drop or
signal, ball type

manually restored drop

electrically restored drop

switchboard-type lamp

indicating lamp

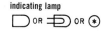 OR   OR (*)

jeweled signal light

INDICATING LIGHT
LETTER
COMBINATIONS

(Replace asterisk (*) with
proper letter combination.)

A amber
B blue
C clear
G green
NE neon
O orange
OP opalescent
P purple
R red
W white
Y yellow

LOGIC SYMBOLS

(Including some duplicate general-purpose symbols; do not mix left-hand symbols and right-hand symbols.)

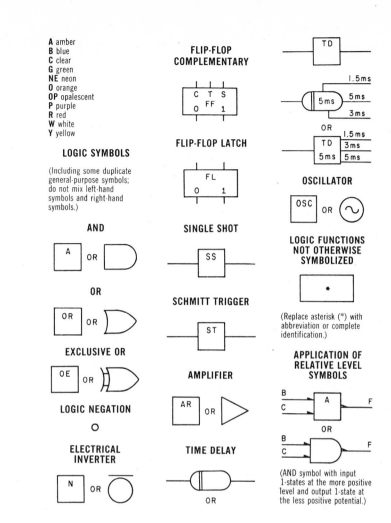

AND

OR

EXCLUSIVE OR

LOGIC NEGATION

ELECTRICAL INVERTER

FLIP-FLOP COMPLEMENTARY

FLIP-FLOP LATCH

SINGLE SHOT

SCHMITT TRIGGER

AMPLIFIER

TIME DELAY

OR

OSCILLATOR

LOGIC FUNCTIONS NOT OTHERWISE SYMBOLIZED

(Replace asterisk (*) with abbreviation or complete identification.)

APPLICATION OF RELATIVE LEVEL SYMBOLS

OR

(AND symbol with input 1-states at the more positive level and output 1-state at the less positive potential.)

ELECTRICAL SYMBOLS FOR ARCHITECTURAL DRAWINGS[1]

1·0 LIGHTING OUTLETS

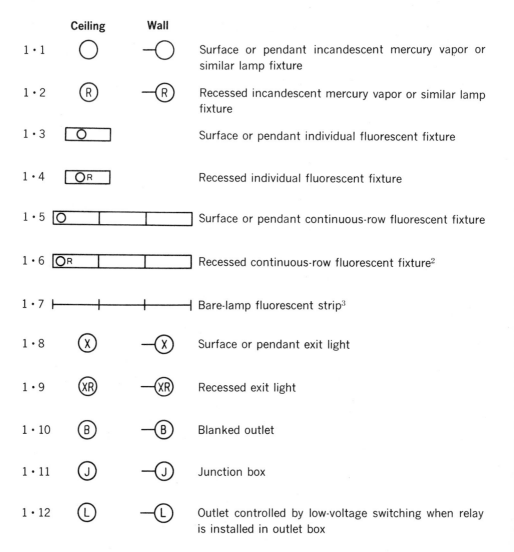

	Ceiling	Wall	
1·1	○	─○	Surface or pendant incandescent mercury vapor or similar lamp fixture
1·2	Ⓡ	─Ⓡ	Recessed incandescent mercury vapor or similar lamp fixture
1·3	▭○▭		Surface or pendant individual fluorescent fixture
1·4	▭○R▭		Recessed individual fluorescent fixture
1·5	○▭▭		Surface or pendant continuous-row fluorescent fixture
1·6	○R▭▭		Recessed continuous-row fluorescent fixture[2]
1·7	├──┼──┤		Bare-lamp fluorescent strip[3]
1·8	Ⓧ	─Ⓧ	Surface or pendant exit light
1·9	ⓍⓇ	─ⓍⓇ	Recessed exit light
1·10	Ⓑ	─Ⓑ	Blanked outlet
1·11	Ⓙ	─Ⓙ	Junction box
1·12	Ⓛ	─Ⓛ	Outlet controlled by low-voltage switching when relay is installed in outlet box

[1]These symbols are taken from ANSI Y32.9–1962, "Graphical Electrical Wiring Symbols for Architectural and Electrical Layout Drawings," published by the American Standards Association and sponsored by the Institute of Electrical and Electronics Engineers and the American Society of Mechanical Engineers.

[2]In the case of combination continuous-row fluorescent and incandescent spotlights, use combinations of the above standard symbols.

[3]In the case of continuous-row bare-lamp fluorescent strip above an area-wide diffusing means, show each fixture run, using the standard symbol; indicate area of diffusing means and type by light shading and/or drawing notation.

2·0 RECEPTACLE OUTLETS

Where all or a majority of receptacles in an installation are to be of the grounding type, the uppercase letter abbreviated notation may be omitted and the types of receptacles required noted in the drawing list of symbols and/or in the specifications. When this is done, any nongrounding receptacles may be so identified by notation at the outlet location.

Where weatherproof, explosion-proof, or other specific types of devices are required, use the type of uppercase subscript letters referred to under Section 0.2, item a-2 of this Standard. For example, weatherproof single or duplex receptacles would have the uppercase subscript letters noted alongside the symbol.

	Ungrounded	Grounding	
2·1			Single receptacle outlet
2·2			Duplex receptacle outlet
2·3			Triplex receptacle outlet
2·4			Quadruplex receptacle outlet
2·5			Duplex receptacle outlet, split wired
2·6			Triplex receptacle outlet, split wired
2·7			Single special-purpose receptacle outlet[1]
2·8			Duplex special-purpose receptacle outlet[1]
2·9	R	RG	Range outlet
2·10	DW	GDW	Special-purpose connection or provision for connection. Use subscript letters to indicate function (DW—dishwasher; CD—clothes dryer, etc.)

[1]Use numeral or letter either within the symbol or as a subscript alongside the symbol keyed to explanation in the drawing list of symbols to indicate type of receptacle or usage.

	Ungrounded	**Grounding**	
2·11	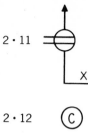		Multi-outlet assembly. (Extend arrows to limit of installation. Use appropriate symbol to indicate type of outlet. Also indicate spacing of outlets as x in.)
2·12	Ⓒ	–Ⓒ ₉	Clock-hanger receptacle
2·13	Ⓕ	–Ⓕ ₉	Fan-hanger receptacle
2·14			Floor single-receptacle outlet
2·15			Floor duplex-receptacle outlet
2·16	⃰	⃰	Floor special-purpose outlet[1]
2·17			Floor telephone outlet, public
2·18			Floor telephone outlet, private

 Not a part of the Standard: Example of the use of several floor outlet symbols to identify a 2-, 3-, or more-gang floor outlet

2·19 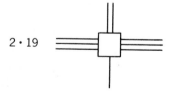 Underfloor duct and junction box for triple-, double-, or single-duct system as indicated by the number of parallel lines

[1]Use numeral keyed to explanation in drawing list of symbols to indicate usage.

Ungrounded   **Grounding**

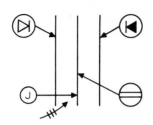

Not a part of the Standard: Example of use of various symbols to identify location of different types of outlets or connections for underfloor duct or cellular floor systems

2·20 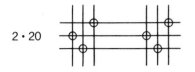   Cellular floor header duct

3·0 SWITCH OUTLETS

3·1   S   Single-pole switch

3·2 S_2 Double-pole switch

3·3 S_3 Three-way switch

3·4 S_4 Four-way switch

3·5 S_K Key-operated switch

3·6 S_P Switch and pilot lamp

3·7 S_L Switch for low-voltage switching system

3·8 S_{LM} Master switch for low-voltage switching system

3·9   S   Switch and single receptacle

3·10   S   Switch and double receptacle

3·11 S_D Door switch

3·12 S_T Time switch

3·13 S_{CB} Circuit-breaker switch

3·14 S_{MC} Momentary contact switch for push button for other than signaling system

3·15   (S)   Ceiling pull switch

4 · 0 INSTITUTIONAL, COMMERCIAL, AND INDUSTRIAL OCCUPANCIES

	Basic symbol	Examples of individual item identification (not a part of the Standard)	

4 · 1

I. Nurse-call-system devices (any type)

Nurses' annunciator (can add a number after it as 24 to indicate number of lamps)

Call station, single cord, pilot light

Call station, double cord, microphone-speaker

Corridor dome light, 1 lamp

Transformer

Any other item on same system: use numbers as required

4 · 2

II. Paging-system devices (any type)

Keyboard

Flush annunciator

Two-face annunciator

Any other item on same system: use numbers as required

	Basic symbol	Examples of individual item identification (not a part of the Standard)

4 · 3

III. **Fire-alarm-system devices (any type) in-**
cluding smoke and sprinkler alarm devices.

 Control panel

⊢2 Station

⊢3 10-in. gong

⊢4 Presignal chime

⊢5 Any other item on same system: use numbers
as required

4 · 9

IX. **Sound system**

⊢◁1 Amplifier

⊢◁2 Microphone

⊢◁3 Interior speaker

⊢◁4 Exterior speaker

 Any other item on same system: use numbers
as required

Basic
symbol

Examples of
individual item
identification
(not a part of
the Standard)

4 · 10

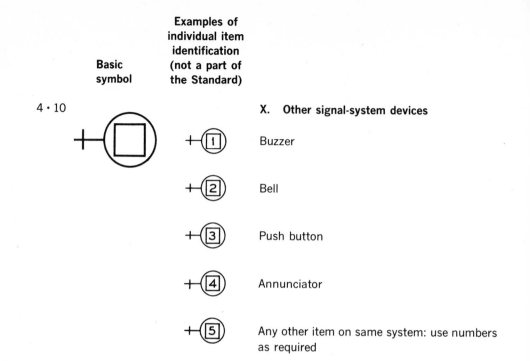

X. Other signal-system devices

Buzzer

Bell

Push button

Annunciator

Any other item on same system: use numbers
as required

SIGNALING SYSTEM OUTLETS

5 · 0 RESIDENTIAL OCCUPANCIES

Signaling-system symbols for use in identifying standardized residential-type signal-system items on residential drawings where a descriptive symbol list is not included on the drawing. When other signal-system items are to be identified, use the above basic symbols for such items together with a descriptive symbol list.

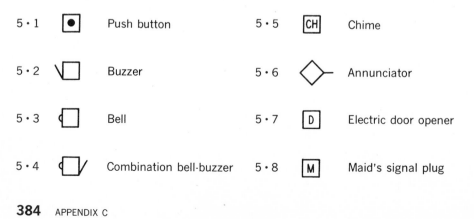

5 · 1　Push button

5 · 2　Buzzer

5 · 3　Bell

5 · 4　Combination bell-buzzer

5 · 5　Chime

5 · 6　Annunciator

5 · 7　Electric door opener

5 · 8　Maid's signal plug

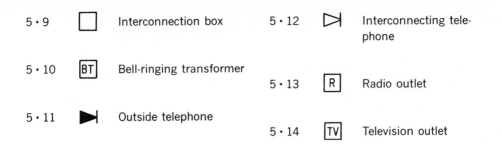

5·9	☐	Interconnection box
5·10	BT	Bell-ringing transformer
5·11	◤	Outside telephone

5·12	◁	Interconnecting tele-phone
5·13	R	Radio outlet
5·14	TV	Television outlet

6·0 PANELBOARDS, SWITCHBOARDS, AND RELATED EQUIPMENT

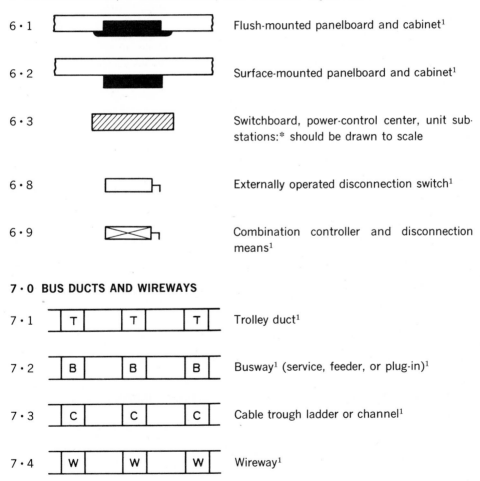

6·1		Flush-mounted panelboard and cabinet[1]
6·2		Surface-mounted panelboard and cabinet[1]
6·3		Switchboard, power-control center, unit sub-stations:* should be drawn to scale
6·8		Externally operated disconnection switch[1]
6·9		Combination controller and disconnection means[1]

7·0 BUS DUCTS AND WIREWAYS

7·1	T T T	Trolley duct[1]
7·2	B B B	Busway[1] (service, feeder, or plug-in)[1]
7·3	C C C	Cable trough ladder or channel[1]
7·4	W W W	Wireway[1]

[1]Identify by notation or schedule.

9·0 CIRCUITING

Wiring-method identification by notation on drawing or in specifications

9·1 ———————————— Wiring concealed in ceiling or wall

9·2 —— —— —— —— Wiring concealed in floor

9·3 – – – – – – – – – – Wiring exposed

Note: Use heavy-weight line to identify service and feeders. Indicate empty conduit by notation CO (conduit only).

9·4 ————————————▶² ¹ Branch circuit home run to panelboard. Number of arrows indicates number of circuits. (A numeral at each arrow may be used to identify circuit number.) **Note:** Any circuit without further identification indicates two-wire circuit. For a greater number of wires, indicate with cross lines, for example: ——///—— 3 wires; ——////—— 4 wires, etc. Unless indicated otherwise, the wire size of the circuit is the minimum size required by the specification.

Identify different functions of wiring system, for example, signaling system by notation or other means.

9·5 O———————————— Wiring turned up

9·6 ————————————● Wiring turned down

10·0 ELECTRIC DISTRIBUTION OR LIGHTING SYSTEM, UNDERGROUND

10·1 ☐M Manhole[1]

10·2 ☐H Handhole[1]

[1]Identify by notation or schedule.

INTERNATIONAL ELECTROTECHNICAL COMMISSION
SYMBOLS FOR CAPACITORS

	Preferred Form	**Other Form**
GENERAL		
If it is necessary to identify the electrodes, the modified element shall represent the outside electrode.		
Nonpolarized electrolytic capacitor		
Polarized electrolytic capacitor		
Polarized semiconductor capacitor		
Polarized semiconductor capacitor if deliberate use is made of the inherent variability		
Lead-through capacitor, feed-through capacitor		
Variable capacitor, general symbol		
Capacitor with preset adjustment		

Variable differential capacitor (**Note:** $C_1 + C_2 =$ constant)

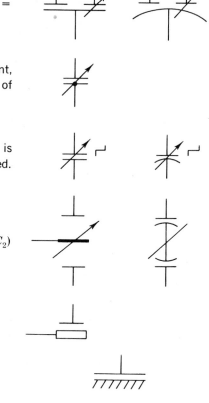

If it is desired to distinguish the moving element, the intersection of the latter with the symbol of variability is marked by a dot.

If it is desired to specify that the capacitor is operating by steps, the symbol for steps is indicated.

Variable split-stator capacitor (**Note:** $C_1 = C_2$)

Capacitor with inherent series resistance

Chassis capacitor

The International Electrotechnical Commission has approved many symbols shown in the American Standard. However, because the recommended capacitor symbols differ considerably from the U.S. symbols, this section of the IEC standard is printed here.

Bibliography

Electrical or electronics drawing

Bishop, Calvin C., C. T. Gilliam, and Associates: *Electrical Drafting and Design,* 3d ed., McGraw-Hill Book Company, New York, 1952.

Carini, L. F. D.: *Drafting for Electronics,* McGraw-Hill Book Company, New York, 1946.

Kuller, K. Karl: *Electronics Drafting,* McGraw-Hill Book Company, New York, 1962.

Mark, D.: *How to Read Schematic Diagrams,* John F. Rider, Publisher, Inc., New York, 1957.

Raskhodoff, N. M.: *Electronic Drafting Handbook,* The Macmillan Company, New York, 1971.

Shiers, G.: *Electronic Drafting,* Prentice-Hall, Inc., Englewood Cliffs, N.J., 1962.

Van Gieson, D. Walter: *Electrical Drafting,* McGraw-Hill Book Company, New York, 1945.

Engineering drawing or graphics

Arnold, J. Norman: *Introductory Graphics,* McGraw-Hill Book Company, New York, 1958.

Black, E. D.: *Graphical Communication,* McGraw-Hill Book Company, New York, 1959.

French, Thomas E., and Carl L. Svensen: *Mechanical Drawing,* 6th ed., McGraw-Hill Book Company, New York, 1957.

———— and Charles J. Vierck: *A Manual of Engineering Drawing for Students and Draftsmen,* 10th ed., McGraw-Hill Book Company, New York, 1971.

———— and ————: *Graphic Science,* 2d ed., McGraw-Hill Book Company, New York, 1963.

General Drafting, TM 5-230, Department of the Army and Air Force, U.S. Government Printing Office, Washington, D.C., 1955.

Giesecke, F. E., A. Mitchell, and H. C. Spencer: *Technical Drawing,* 5th ed., The Macmillan Company, New York, 1967.

Hoelscher, R. P., C. H. Springer, and J. Dabrovolney: *Engineering Drawing and Geometry,* 3d ed., John Wiley & Sons, Inc., New York, 1961.

Levins, A. S.: *Graphics with an Introduction to Conceptual Design,* John Wiley & Sons, Inc., New York, 1962.

Lombardo, J. V., L. O. Johnson, and W. I. Short: *Engineering Drawing,* Barnes & Noble, Inc., New York, 1953.

Luzadder, W. J.: *Fundamentals of Engineering Drawing,* Prentice-Hall, Inc., Englewood Cliffs, N.J., 1959.

————: *Basic Graphics,* 2d ed., Prentice-Hall, Inc., Englewood Cliffs, N.J., 1968.

Mochel, M. G.: *Fundamentals of Engineering Graphics,* Prentice-Hall, Inc., Englewood Cliffs, N.J., 1960.

Orth, H. D., R. R. Worsencroft, and H. B. Doke: *Theory and Practice of Engineering Drawing,* 2d ed., William C. Brown Company, Dubuque, Iowa, 1959.

Paré, E. G.: *Engineering Drawing,* Holt, Rinehart and Winston, Inc., New York, 1959.

Rising, J. S., and M. W. Almfeldt: *Engineering Graphics,* 3d ed., William C. Brown Company, Dubuque, Iowa, 1964.

Zozzora, Frank: *Engineering Drawing,* 2d ed., McGraw-Hill Book Company, New York, 1958.

Miscellaneous, electronics or electrical

Allis-Chalmers: *Motor Control Theory and Practice,* Allis-Chalmers Manufacturing Company, Milwaukee, 1961.

Bureau of Naval Personnel: *Aviation Electronics Technician 3 & 2,* U.S. Government Printing Office: Washington, D.C., 1959.

Cage, J. M.: *Theory and Application of Industrial Electronics,* McGraw-Hill Book Company, New York, 1951.

Carroll, John M.: *Electron Devices and Circuits,* McGraw-Hill Book Company, New York, 1962.

Chute, George M.: *Electronics in Industry,* 3d ed., McGraw-Hill Book Company, New York, 1964.

G.E. Transistor Manual, General Electric Semiconductor Products Department, Syracuse, N.Y., 1962.

Gillie, Angelo C.: *Pulse and Logic Circuits,* McGraw-Hill Book Company, New York, 1968.

Goldman, Richard: *Ultrasonic Technology,* Reinhold Publishing Corporation, New York, 1962.

Heumann, G. M.: *Magnetic Control of Industrial Motors,* John Wiley & Sons, Inc., New York, 1954.

Hill, W. Ryland: *Electronics in Engineering,* 2d ed., McGraw-Hill Book Company, New York, 1961.

Keonjian, Edward: *Microelectronics,* McGraw-Hill Book Company, New York, 1963.

Kiver, Milton S.: *Transistors,* 3d ed., McGraw-Hill Book Company, New York, 1962.

Malvino, A. P., and Leach, D. P.: *Digital Principles and Applications,* McGraw-Hill Book Company, New York, 1969.

Markus, John, and Vin Zeluff: *Handbook of Industrial Electronic Control Circuits,* McGraw-Hill Book Company, New York, 1956.

"National Electrical Code," American National Standards Institute, New York, 1971.

Phillips, Alvin B.: *Transistor Engineering,* McGraw-Hill Book Company, New York, 1962.

Radio Corporation of America: *Transistor Manual,* RCA Semiconductor and Materials Division, Somerville, N.J., 1962.

Schmid, C.F.: *Graphic Presentation,* The Ronald Press Company, New York, 1954.

Slurzberg, Morris, and William Osterheld: *Essentials of Electricity for Radio and Television,* McGraw-Hill Book Company, New York, 1950.

Index